THE
CELL

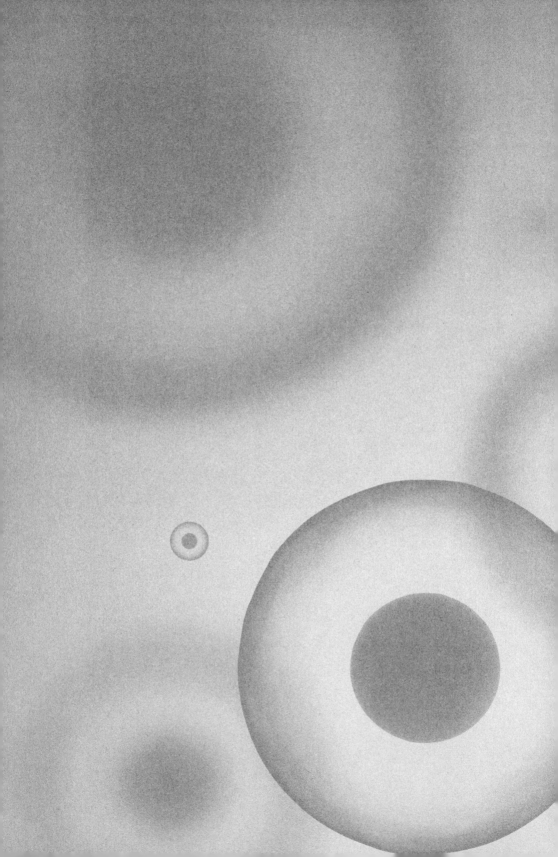

THE
CELL

DISCOVERING THE MICROSCOPIC WORLD THAT DETERMINES OUR HEALTH, OUR CONSCIOUSNESS, AND OUR FUTURE

JOSHUA Z. RAPPOPORT, PhD

BenBella

BenBella Books, Inc.
Dallas, TX

BenBella Books, Inc.
10440 N. Central Expressway, Suite 800 | Dallas, TX 75231
www.benbellabooks.com
Send feedback to feedback@benbellabooks.com

Printed in the United States of America
10 9 8 7 6 5 4 3 2 1

Library of Congress Cataloging-in-Publication Data
Names: Rappoport, Joshua Z., 1974-
Title: The cell : discovering the microscopic world that determines our
    health, our consciousness, and our future / Joshua Z. Rappoport, PhD.
Description: Dallas, TX : BenBella Books, Inc., [2017] | Includes
    bibliographical references and index.
Identifiers: LCCN 2016046562 (print) | LCCN 2016053541 (ebook) | ISBN
    9781942952961 (paperback) | ISBN 9781944648978 (electronic)
Subjects: LCSH: Cytogenetics. | BISAC: SCIENCE / Life Sciences / Cytology. |
    SCIENCE / Applied Sciences. | MEDICAL / Diseases.
Classification: LCC QH441.5 .R37 2017 (print) | LCC QH441.5 (ebook) | DDC
    572.8—dc23
LC record available at https://lccn.loc.gov/2016046562

Editing by David Bessmer
Copyediting by Scott Calamar
Proofreading by James Fraleigh and Chris Gage
Indexing by Debra Bowman
Text design and composition by Silver Feather Design
Front cover design by Brian Barth
Full cover design by Sarah Dombrowsky
Printed by Lake Book Manufacturing

Distributed by Perseus Distribution
www.perseusdistribution.com
To place orders through Perseus Distribution:
Tel: (800) 343-4499
Fax: (800) 351-5073
E-mail: orderentry@perseusbooks.com

Special discounts for bulk sales (minimum of 25 copies) are available.
Please contact Aida Herrera at aida@benbellabooks.com.

Throughout each stage of scientific training, one gets to work alongside more experienced supervisors who lead us, for better or worse, by example. I have been extremely lucky to have the pleasure of working with a series of unbelievably capable, giving, and thoughtful mentors throughout the last twenty-five years.

While I was an undergraduate at Brown University, I took several classes with Professor Don Jackson and did a series of research projects in his lab. A first-class empiricist, Dr. Jackson instilled in me an appreciation of the order inherent in the natural world and taught me that every repetition of an experiment must be performed absolutely identically. My PhD supervisor at Mount Sinai School of Medicine was Professor Ruth Abramson, who, sadly, passed away in 2004. Dr. Abramson's work ethic and moral fiber served as an example not only to those lucky enough to be among her students, but to all who knew her.

After my PhD, I had the honor and privilege to work with Professor Sanford Simon at the Rockefeller University. More than anyone else, Sandy is responsible, for good or ill, for the scientist that I am today. Sandy's laser-sharp focus, genuine excitement for science, and excellent sense of humor—combined with his absolutely rigorous focus on quantification and statistical analysis and the importance he places on the happiness of the people working in his lab—shaped me in numerous ways. Sandy sees that everyone is an individual who can't be expected to fit some unrealistic ideal mold. He recognizes that key contributions can be made equally by both the high school–student intern and the visiting professor. The perspective that each scientist is a person, not a cog in a machine, and that a lab works best when people get along together, rather than compete for the boss's attention or the lead role in the next big project, is sorely lacking in science today, but is exemplified by Sandy.

These people, these excellent scientists and caring educators, gave of their time to help train me, and for that I am eternally grateful. Science involves the ability to read the literature critically, to employ reasoning and logic as well as technical skill. Each of these three mentors worked very hard to help me develop these attributes. That is why I dedicate this book to Sandy Simon, Don Jackson, and the memory of Ruth Abramson, each of whom contributed to my education and growth as a scientist and a person.

When you employ the microscope, shake off all prejudice, nor harbor any favorite opinions; for, if you do, 'tis not unlikely fancy will betray you into error, and make you see what you wish to see.

—HENRY BAKER, 1742

Contents

PART SIX
The Profession of Cell Biology—
The Good, the Bad, and the Future

Why Read This Book?

The world of biomedical research moves at breakneck speed. It seems as if we hear about new breakthroughs almost daily. Everyone cares about new discoveries, because they affect our health and longevity. Moreover, this kind of news not only has to do with causes, diagnosis, and treatment of disease, but sometimes suggests ways in which we should change our behavior, diets, and lifestyles. In fact, it's those latter items—what you can do to live longer and healthier—that tend to be most sensationalized, exaggerated, and misunderstood.

Whether you are talking about obesity, aging, infectious diseases, cancer, or improving your Sudoku abilities, it all comes down to cells and the molecules they create and that regulate them. Thus, one of my purposes for writing this book is to give the average person a better understanding of the basic science of cellular and molecular biology so that you can evaluate more critically what you hear and read. Another is to lead you through one of the most thrilling and fascinating modern frontiers of human discovery, in particular from the perspective of a microscopist armed with the most powerful means to visualize events at the cellular and molecular scale.

Understanding science isn't always easy, and neither is communicating it accurately to the general public. Universities, hospitals, and research institutes now have media relations departments that issue newsletters, press releases, and tweets announcing the latest, greatest discoveries from the researchers they employ, so that we are constantly bombarded by abstractions intended to convey sensational scientific breakthroughs in as few and simple words as possible. Meanwhile, scientists now need to be skilled at delivering brief, engrossing, and pithy elevator pitches and "community engagement" in order to convey their research goals and experimental results to nonscientists quickly and clearly.

Popular media, which is in the business of retailing the sensational, tends to express breathless excitement rather than accurate scientific information. We generally hear not about what the actual research in question showed, but what it could lead to. Most science writers are primarily journalists who can't be blamed for occasional inaccuracies or for focusing more on the potential ramifications of new advances, rather than the precise description of details. Conversely, most scientists are, quite frankly, terrible communicators. It is a rare researcher who can describe his or her work to nonscientists accurately and include the significant information without putting the listener to sleep.

Who am I to think I can do any better?

I don't want to convince you that my specific area of research, my particular sector of the scientific universe, is the most fascinating and relevant. Rather, I want to provide a basic foundation and framework for discussing a wide variety of interesting, important, and timely topics in biomedical research.

My perspective as a cell biologist is that cells underlie most of what goes wrong in disease. Cells are the targets of many therapies, historical and cutting edge. As a microscopist, it seems to me that much of what we know about the inner workings of biological systems comes from direct visualization with microscopes. We are currently in a golden age of microscopy. Things that were previously hidden from view are now readily visible, and microscopes now serve not only for the observation of molecules, cells, tissues, and organisms, but as experimental tools permitting the manipulation and interrogation of biomedical systems.

Thus, my goal is not to describe my own research, or to offer a systematic and comprehensive textbook. I just want to convey interesting and scientifically accurate information on a wide range of interrelated topics, setting the stage for a greater understanding of the most important biomedical challenges we face as a species. It is my hope that this book will serve as a starting point toward a greater understanding of biology and medicine and the advances that affect our health and well-being.

There is another very important reason why I wanted to write this book. Although anti-intellectualism has always existed, today's twenty-four-hour news cycle, sharp cultural divisions, and the rise of Internet trolls have sadly made "antiscience" a prominent factor in our national conversation. While skepticism lies at the heart of the scientific method and the self-correcting nature of scientific research, things like faith, belief, and gut feelings, although important in other contexts, are not valid counterarguments to scientific conclusions and consensus. Whether the question is evolution, climate change, or the safety of vaccines, too much attention—not to mention credibility—is given to "alternatives" to accepted, scientifically proven perspectives. The penchant for giving equal attention to false equivalences without demanding empirically determined scientific facts is illogical, unwise, and dangerous.

I hope that by reading this book you will get an insider's view into science and the work of scientists who have made particularly relevant and interesting observations. The scientific method involves identifying a question, a hypothesis, and then devising some sort of experiment to test it and thereby further our understanding of the natural world. The results of experiments must be quantified and then statistically tested. In some cases, such as deducing the structure of a molecule—for example, the DNA double helix—a distinct hypothesis is really not being tested per se. But when a hypothesis is subjected to experiment, the methodology employed and data acquired are made available for other scientists to judge and to reproduce in their own laboratories. It is certainly true that scientists can get things wrong, data can be misinterpreted, and errors in experiment design and methodology can be made. But ultimately, over the years as many scientists try to answer the same types of questions, the correct answer emerges and is confirmed. Frauds are exposed and mistakes are corrected.

Scientists do not engage in large-scale conspiracies. In fact, vast networks of nefarious scientists working in different places around the world *do not exist*, because you can't get that many people to keep *anything* a secret, much less skulduggery—especially when many of those people are in competition with one another. Scientists do strive for funding and work in competitive environments. Success is far from ensured, and pressure to make your mark is constant. So, while a few cases of hidden malfeasance may not come to light, at least not immediately, the competitive nature of scientific endeavor ensures that the most interesting and important results cannot be faked, at least not for long. Scientists love nothing more than proving other scientists wrong; only a shortsighted and immature scientist could think that faked results will not ultimately be exposed.

Contrary to the clamor of some anti-intellectuals, we scientists are not in the pockets of pharmaceutical companies, nor are we holding back discoveries or deliberately engaging in mass conspiracies for wicked or self-serving purpose. First of all, we generally don't get paid enough to rationalize that kind of unethical and illegal behavior. We have way too much to lose and too little to gain. When scientists are shown to have faked data, their work is expunged from the scientific record, and they must bear the shame of having a paper retracted from a journal. Even when tenured, they can lose their jobs and their careers. On the other hand, when honest scientists are proven wrong in light of newer data, when further information shows that they had the wrong end of the stick on something, they shrug their shoulders, accepting that that is how progress goes, and get back to work.

I hope that by reading this book, you will gain insight into how science *really* works and thus why science should be promoted and supported, rather than treated as some impenetrable and mysterious dark art that can be bent insidiously to political and economic goals.

I am a cell biologist and microscopist. Over the course of the last twenty years or so, I have spent an inordinate amount of time using microscopes to look at cells in the hope of understanding various fundamental aspects of life on the cellular scale. I have studied how cellular waste products are eliminated, how the cell takes in substances from the outside space, and how cells move from one place to another—for example, during wound healing or cancer metastasis. Furthermore, I have been involved in developing microscope-based experimental approaches. This has left me with a firm belief that by using microscopes to study cells we can learn a tremendous amount about, to lift a phrase from Douglas Adams, author of *The Hitchhiker's Guide to the Galaxy*, "life, the universe, and everything."

Why cells?

The cell is the fundamental unit of life. Single-celled organisms cover every surface of our world, including our bodies. The human body is an interconnected arrangement of different self-contained organs consisting of specific cell types arranged into particular configurations. Our cells can be dissociated and grown in a dish. Diseases such as cancer can start as single malfunctioning cells. Cellular therapies, the potential of stem cells, and much of modern personalized and regenerative medicine all boil down to new ways to analyze, understand, and manipulate cells. Without an understanding of cells and cell biology, there is no way to comprehend modern biomedical research and clinical practice. Thus, this book looks at cells as the central focus of human health and disease, the inner workings of our bodies, and the main therapeutic targets of modern medicine.

Why microscopy?

I have spent a great deal of my waking life staring down microscopes, sitting in dark rooms patiently observing life at the finest levels of scale possible. In my lifetime, the development of microscopy technology and techniques has accelerated incredibly. The last few decades have seen the implementation of new types of microscopes that would be as unrecognizable to the founding pioneers in this field as supercomputers would be to people like Alan Turing, or maybe even the ancient Mesopotamians who developed the first abacus. Until very recently, light microscopes were limited in resolution to the scale of a micron, or, at the very best, about a quarter of a micron. A micron is one millionth of a meter; an average human hair is about 100 microns thick. A nanometer (nm) is one thousandth of a micron, about the size of four water molecules lined up together. Due to very recent technological advancements, we can now routinely study cellular structures at the nano scale in live cells. Researchers around the world can now image the structures and compartments inside intact living cells at a resolution more than ten times greater than what was previously thought to be a physically imposed limit. So what we are doing now isn't really microscopy anymore, but "nanoscopy."

This is a massive shift in our ability to interrogate biological systems and to study and understand life at the cellular scale. Most cells are somewhere between 10 and 100 microns across; we've been able to see them through microscopes since the seventeenth century. What the new techniques give us is the ability to image cells' components, many of which are on the order of 10 to 100 nanometers across and are very tightly packed together inside the cell. Thus, this increased resolution is already significantly improving the ways we can visualize and understand the fundamental processes of life.

Resolution is not the same as magnification. If you zoom in on (i.e., magnify) a digital photo on your computer screen, the resolution becomes gradually worse until all you can see is a few large pixels. However, the higher the resolution—i.e., the more pixels per square inch—the more detail you can see and the closer you can zoom in before the image becomes a pixelated blur. In microscopy, the inherent resolution of your instrument is its ability to visually distinguish separate, adjacent objects, even before an image is acquired with a camera. This is a quality of a microscope system that is related to, but not the same as, the degree of magnification. Because of light's wavelike properties, you can only really discern two adjacent objects that are physically separated by a distance greater than about half the wavelength of the light that you are using to illuminate them; otherwise the light waves effectively combine, blurring the distinction between the two objects.

A rainbow shows us the different colors that constitute the spectrum of visible light. These different colors represent particular ranges of wavelength. The human eye can only detect light waves between about 400 nm and 700 nm. Similarly, the glass that microscope lenses and filters are made from is unable to transmit ultraviolet light below the bottom end of this range efficiently. Using light in the visible range, the best we can do is to distinguish objects that are at least 250 nm apart from each other. Unfortunately, the cell is a very busy and crowded place where most things are packed closer together than this resolution limit.

For analogy, imagine that you are a visitor from another planet floating around in your spaceship in high orbit above the Earth. It is your job to try to understand humans based on the images you can acquire. If the best your technology could do was to image an individual human only if they were at least ten feet away from someone or something else, you would have a really difficult time. You would probably end up thinking that humans spent most of their time hiking, jogging, and swimming, because those would be the only times you could actually distinguish a human being from things like trees, cars, buildings, and other humans. At least you could study humans in groups, going on pub crawls and gathered in stadiums.

Imagine that you could label people by dipping them in DayGlo paint so that you could at least differentiate them from their surroundings. You would be able to see the person walking down the street or driving in a convertible, but as soon as more than one person was present within that ten-foot radius, you would have no idea if you were studying one, ten, or a hundred people. That is the trouble with light microscopy. The things we image are generally much smaller and, more important, closer together than the resolution limit. If you are studying the synaptic vesicles that carry neurotransmitters in the brain, which can be less than 50 nm across and very densely packed in the neurons you are imaging, how could you possibly tell how many exist within a particular space if you could only resolve things more than 250 nm apart?

Now we can do that.

Electron microscopes, which provide excellent resolution, have been around for nearly a hundred years. Since electrons are much smaller than photons, resolution of even a few nanometers is possible. However, using electron microscopy to efficiently and convincingly label and image specific substances or compartments in the cell is extremely time-consuming. You have to take and assemble a huge number of tiny pictures. Moreover, it's only possible in cells that have been chemically treated—for example, imbedded in plastic and cut into very thin slices for imaging, which is to say: dead. Fluorescence microscopy uses as beacons molecular tags called fluorophores that absorb light of one color and emit light of another, allowing us to mark the location of any cellular protein, organelle, or even chemical reaction. Now, with nanoscopy (or super-resolved fluorescent microscopy, as the Nobel Prize announcement in 2014 called it), we can finally see what is going on at the subcellular scale. That is, we can image with molecular detail in *living* cells.

This is truly a transformative technology with great promise for the future of cell biology. Understanding the molecular basis of disease is now truly within our grasp, and the collaboration of physicists, engineers, chemists, biologists, and computer scientists is permitting analyses at levels of scale that, until very recently, were thought to be impossible. We now routinely conduct fluorescence microscopy at more than ten times the resolution previously thought to be the maximum possible. The current record for a fluorescence microscope is a resolution of approximately 20 nanometers, and this can be done with a commercially available system that is about as difficult to master as connecting your phone to a rental car via Bluetooth.

Where do we go from here? The coming years will see fits and starts with new techniques and technologies emerging, some appearing and falling away like shooting stars, others here to stay. Already the pages of scientific journals are filling with images that look very different from those published just a few years ago. Nearly all biomedical researchers now have access to this technology, and it is not the only revolutionary microscopy-based approach currently being developed and implemented.

We can specifically label molecules and compartments in living cells, and we can track individual cells in living animals. We can watch metastatic cancer cells exit a tumor in search of a site to create a new secondary tumor. We now have the ability to continuously image and track every cell in a living, developing worm or fruit fly without causing any damage to the specimen from constant illumination with laser light. Single-plane illumination microscopy (SPIM), also called light-sheet imaging, is not a new concept, but it is now a commercial reality that offers very-high-resolution imaging of what is happening inside intact, living organisms without causing any damage. We also have the ability to use new types of optics to "see" all the way through the brains of larger model organisms such as mice, and this can be combined with techniques that allow us to visualize the firing of individual neurons, like imaging lightning bolts in real time. We can identify synapses formed between any two types of neurons we want to study, and we can mark wherever two proteins bind to each other or come in very close proximity. We can measure the flow of blood through the smallest capillaries and study the formation of blood clots or atherosclerotic plaques *as these complex multicellular events unfold* in a live animal. As with the x-ray machine, the MRI, and the sonogram, the ability to better visualize organisms, organs, tissues, cells, and structures of interest has always engendered immediate, tremendous leaps in our understanding.

The last ten years have seen more development in our ability to image cells than the 350 years prior. And that brings us to the real beginning of this story: in 1665, in London, at the Royal Society, with a thirty-year-old clergyman's son named Robert Hooke.

PART ONE

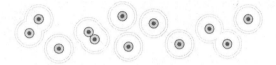

WHAT CELLS ARE
AND HOW THEY WORK

CHAPTER I

A Day the World Changed

◉

O NE DAY IN THE 1660S, A BRILLIANT BRITISH "NATURAL philosopher" named Robert Hooke peered into a new kind of optical instrument and saw something no human had ever seen before. He was using this device, a microscope, to examine a thin slice of cork backlit by an oil lamp. What he saw changed human understanding of the world in a way at least as profound as the first telescopic examination of the planets and their moons by Galileo and others earlier in that century. He saw that cork was constructed of a huge number of tiny, apparently hollow spaces separated by shared walls.

Hooke called these spaces cells, because they reminded him of the little rooms in which monks slept, and that is what we still call them today. This was a critical watershed moment in human understanding. For the first time, humans grasped that living matter wasn't composed of unique, irreducible, solid materials; rather, there existed a previously invisible world where substances such as wood, leaves, fruit pulp, muscle, liver, skin, and so on were living tissues constructed of the same kind of basic components—cells.

These two men, Galileo looking outward through his telescope at the enormity of the solar system and the universe beyond, and Hooke looking inward at the first suggestion of the vastness of the microscopic world, were among the founders of science as we know it.

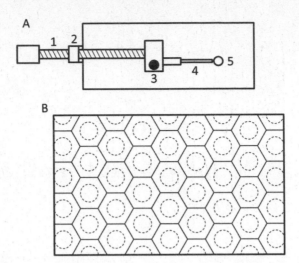

A.   Diagram of a Leeuwenhoek-type microscope. (1) Screw
to move the sample. (2) Connection between screw and
microscope. (3) Focus knob. (4) Needle upon which to place
sample. (5) Objective lens. The length of the microscope
could be as small as a few inches.

B.   Drawing of what a confluent monolayer of cells might
look like through a microscope. Dotted circles are the cell
nuclei, and the hexagons are the cell borders.

Hooke's discovery of the cell would lead to our current understanding of how organisms begin life; how they function, reproduce, and
die; and how species evolve by adapting to changing environments. In
their era, the beginning of the Enlightenment, science—observation,
experimentation, and reason—began to replace magic and mythol-
ogy as the basis for understanding the physical world. Furthermore,
while the heliocentric worldview of Galileo, Copernicus, and Kepler
removed humans and our humble rocky home from the center of
the universe, the work of Hooke and others like him began to show,
nearly 200 years before Darwin published *On the Origin of Species*,
that all living matter shares common traits—in particular, a structural
organization based upon the fundamental unit of the cell.

In 1665, Hooke published a phenomenally popular book, *Micrographia*, that showed the general public what he had been seeing through his microscope. In addition to amazingly detailed drawings of insects and plants, the book contained the first description of the cell. Oxford educated, Hooke was the Curator of Experiments for the Royal Society, a group of scientific gentlemen at whose weekly meetings in London he would demonstrate cutting-edge experiments in many different fields. Like many of the scientific luminaries of his time, Hooke was not tied to one specific area of inquiry. He invented the air pump, crosshairs for telescopes, and the wheel barometer, and contributed to the theory of gravity, which led to a lifelong dispute with a fellow member of the Royal Society named Isaac Newton. Hooke helped plan the rebuilding of London after the Great Fire of 1666. A physical law of elasticity is named after him, the extremely succinct and elegant "extension is proportional to force." As noted, Hooke was a superb illustrator and draftsman, and today he is perhaps most widely known for the illustrations in *Micrographia*, which depicted brilliantly a previously unknown and unseen world that exists all around us.

Hooke was the first person to publish a description of a microbe, a living organism not otherwise visible to the naked eye. Upon looking at a particularly furry type of mold, Hooke produced a fascinating description of the fungus Mucor. He realized that mold was basically "microscopical mushrooms," which he referred to as having a "vegetative body," a term still used today, with "stalks" holding up what he surmised might be "seed cases." Unfortunately though, this was just a supposition, as Hooke's microscope was not able to resolve the tiny spores that were inside of these structures, now called sporangia.

Polymath though he was, Hooke did not build his own microscopes. They were provided by instrument maker Christopher Cock. Hooke helped develop many refinements and improvements to Cock's microscopes, such as the use of indirect, reflected light. The microscopes Hooke used looked a great deal like telescopes. The use of multiple lenses—the compound microscope—allowed for high magnification but caused significant aberrations that made it difficult to clearly visualize very small objects and structures. For all his fame and ingenuity, Hooke would never be able to see things as well as a relatively uneducated linen draper who, toiling in obscurity only two hundred miles away in the Netherlands, was making his own much simpler microscopes to astounding effect.

Antoni van Leeuwenhoek spoke no English, but with the help of neighbors in the city of Delft, who served as translators, he began a correspondence with the Royal Society soon after the publication of Hooke's seminal work. It has been suggested that his interest in microscopy began as a result of having to use weak magnifying glasses to look at fabric, but whatever the motive, Leeuwenhoek developed the means to create small glass lenses of surprisingly high quality, which he was then able to polish to pristine clarity. The simple but powerful single-lens microscopes that he designed and built were capable of magnifications of up to 250 times. He was thus the first person to see bacteria, red blood cells, sperm, the process of fertilization, and insects hatching from eggs. Leeuwenhoek also devised an ingenious way to enclose samples in glass capillary tubes, a technique still employed today for certain types of microscopy studies.

Unlike Hooke, Leeuwenhoek never published a book, but he continued to correspond with the Royal Society, which published his letters. However, Leeuwenhoek likely had a copy of Hooke's *Micrographia*, and in his own work he sought to correct what he saw as a critical error of Hooke's, namely the number of segments present in the horn of a louse (although it is possible that the two had looked at lice of different species or developmental stages). Also, Leeuwenhoek was notoriously secretive about his methods. In 1685, he went over a few of his past observations with a visiting Irish physician named Thomas Molyneux, who wrote a letter about the encounter to the Royal Society. Surprisingly, Leeuwenhoek declined to demonstrate the best of his microscopes, and the young observer was left in the dark regarding Leeuwenhoek's most-secret techniques and designs.

From a modern viewpoint, we might consider Hooke more a pure scientist and Leeuwenhoek more a technician or engineer, but the comparison underscores a critical theme in the history of cell biology—in Hooke's words (from the preface to *Micrographia*), "Microscopes producing new Worlds and Terra-Incognita's to our view." To this day, our ability to understand what goes on inside cells depends in large part on advancements in technology that enable us to see smaller things better.

However, true cell biology is not simply the cataloging of different types of cells, tissues, and organisms; otherwise, cell biology would just be an extension of anatomy. Rather, understanding the inner workings of the cell, opening up the still-ticking watch to reveal its varied parts moving in synchrony, is the real calling of the cell biologist. In fact, over the past twenty years or so, many departments of cell biology and anatomy have recast themselves as departments of molecular cell biology. This change in focus demonstrates that the goal of the modern cell biologist is not mere description, but to place the fundamental molecular events underlying life's most basic structures and functions within the context of the living cells. This is where the power of modern microscopes comes to play, not just in permitting a broad surface description of different types of cells, but by truly allowing us to study life at the molecular scale in a living system—a level of observation Hooke and Leeuwenhoek could never have dreamed of, a true synthesis of molecular biology and cell biology.

However, before we embark on that adventure, it will be helpful to many readers if we do, in fact, begin with a tour of the anatomy of the cell. For many, this will be a brief refresher course—a review of what you learned in high school or college—so that you'll be ready for everything that follows.

CHAPTER 2

A Guided Tour of the Cell

C ELLS ARE THE FUNDAMENTAL SUBUNITS OF LIFE. AS ATOMS are to matter, cells are to life. All living things are made of cells.

The first living organisms were single cells, and there are now many species of single-celled organisms. In fact, pretty much everything on Earth, including you, is covered with microscopic bacterial cells. On the other hand, animals such as humans are made up of many kinds of specialized cells—lung cells, skin cells, nerve cells, muscle cells—all with distinct identities and functions. How big is a cell? That depends on what kind of cell we're talking about. Sizes vary far more than you might think. Every time you make an omelet, you are seeing cells with the naked eye. That said, it is the germinal disk, a tiny spot on the side of the egg yolk, that contains the cell nucleus and divides to form the embryo. The yolk is simply a nutrient source. Furthermore, eggs aren't technically cells because they only have half the total amount of DNA required for life. Only a fertilized egg is a complete cell, and the egg industry prevents fertilization from happening, ensuring you don't find a little chick in your frying pan. At any rate, yes, there are cells that can be seen by the naked eye, and not just hens' eggs. There are individual nerve cells in the body that run from your spinal cord all the way to your big toe that can be seen with

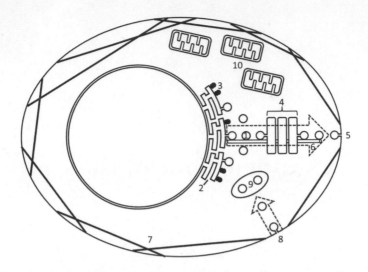

A diagram of the cell and its processes: Cargo containers called secretory vesicles (1) are transported from the endoplasmic reticulum (2), loaded with proteins synthesized by ribosomes (3), through the Golgi apparatus (4), and then fused with the plasma membrane (5), making their cargo available for the rest of the body. Microtubules (6) serve as tracks along which cargo-containing vesicles are transported, while the actin cytoskeleton (7) provides support to the plasma membrane. Following endocytosis (8), endocytic vesicles fuse with multi-vesicular bodies (9), which fuse with lysosomes for protein degradation. The mitochondria (10) are the main site of energy metabolism.

the naked eye. (That would require the help of a surgeon, though, so you should probably take my word for it.) However, these are exceptions. At the other end of the scale, there are many cells in our bodies that we can only see with a very powerful microscope.

It will probably help you to get a sense of the scale of things that we'll be discussing shortly. For those who aren't up on microscopic units of measure, or the metric system, let's pause for a quick review.

1 meter =			
100 centimeters	1 centimeter =		
1,000 millimeters	10 millimeters	1 millimeter =	
1 million microns	10,000 microns	1,000 microns	1 micron =
1 billion nanometers	10 million nanometers	1 million nanometers	1,000 nanometers

To be visible without magnification, a dark spot on a light background has to be at least 100 microns across. That's 1/10th of a millimeter, and 1/100th of a centimeter. The smallest cells in our bodies, such as certain kinds of blood cells, immune cells (leukocytes), and brain cells (granule cells) are less than 10 microns in diameter. That means you would never be able to see them with the unaided eye.

Bacteria and viruses are much smaller yet. Bacteria are simple cells that lack many of the internal compartments or organelles that carry out different biochemical functions in human body cells. Most people think of bacteria as "germs" that make us sick; anthrax, bubonic plague, syphilis, tuberculosis, and salmonella, for example, are all bacterial diseases. However, there are many helpful and beneficial bacteria. We hear in the news about people getting sick from *E. coli*, but there are other, nonpathogenic forms of *E. coli* normally found in the digestive tract that are necessary to help us digest food. We would get sick without them, which sometimes happens when heavy doses of antibiotics reduce the levels of these helpful bacteria along with the bad ones.

Bacteria come in many shapes and sizes; chains and colonies of bacteria can be visible to the naked eye. However, an individual *E. coli* cell is only about half a micron by 2 microns in size.

Viruses are smaller and still less complex than bacteria, and in fact cannot be thought of as cells. These tiny molecular assemblages can only reproduce with the help of a host cell. In fact, even though viruses have genomes and do indeed reproduce, scientists argue about whether they should be thought of as living.

Bacteriophage (or simply phage) are very simple viruses that infect bacteria, and can be less than 100 nanometers long. Phage, which look like little lunar-landing modules, attach to much larger bacterial cells and deposit their genome into the host, turning the bacteria into little factories making more phage, which can eventually burst out of the host and further their own life cycle. Even the viruses that infect humans are generally extremely small. For example, the human immunodeficiency virus (HIV), which causes AIDS, is only about 120 nm across.

So you can see that our guided tour of the cell is going to require a very small tour bus, as it were. Our little vehicle will have to be much smaller than the submarine that traveled around in a patient's bloodstream in that old science fiction movie *Fantastic Voyage*. But before we start, we should probably take a look at a tourist map of the cell we'll be traveling around in, the one shown at the start of this section, to get a general idea of what we'll be seeing. Don't worry about what all the labeled structures are—we'll cover all that as we go.

Let's start our tour. As we approach the cell, we see that it has an outer boundary—this is referred to as the cell membrane or plasma membrane. The latter term is used because it is what separates the cell from the plasma, the fluid component of blood. This is not, however, an impermeable barrier. Obviously, various kinds of substances have to enter the cell—nutrients that the cell needs in order to survive and grow, dissolved salts and sugars, and signals from other cells that regulate specific cellular functions. Likewise, stuff needs to get out, such as the neurotransmitters that relay signals from one nerve cell to another, various hormones, and other signaling molecules, and indigestible junk that the cell needs to eject.

The plasma membrane is a two-sided sheet of lipid molecules that regulates access of proteins, ions, and other molecules into and out of the cell. There are various compartments inside the cell—organelles—that are also bounded by membrane bilayers, including the nucleus, where DNA resides. The space between the plasma membrane and the nucleus is filled with a fluid called cytosol. There are many proteins within organelles, floating free in the cytosol, and residing within the plasma membrane. These proteins serve many different functions: the building and repair of structural components; moving signals into, out of, and within the cell to set various functions in motion; the machinery that synthesizes new genetic material; and the anchors that hold cells in place and, when rearranged, allow cells to move from one place to another. Enzymes are proteins that bring about energy-dependent chemical changes. Integral membrane proteins reside within lipid bilayers; they are partly imbedded within the membrane, but generally have peripheral domains that extend away from the membrane on either side. Integral membrane proteins serve variously as the pumps, transporters, and channels that selectively move molecules from one side of the membrane to the other; they are both gates and gatekeepers. Some also function as receptors for nutrients and extracellular signals.

Many of these receptors reside within the plasma membrane, waiting for a specific, matching type of extracellular molecule to float by and bind to them, like a key fitting into a lock. When this happens, the receptors transmit that signal into the cytosol, for example by activating an enzyme, a protein that accomplishes some function or does some work. These signals might also be growth factors that tell the cell to start the process of cell division. (It is no surprise that too much of this signaling, for example from a mutant receptor that functions in the absence of the extracellular activator, can result in the formation of cancer by inciting inappropriate and uncontrolled cell division.)

How integral membrane proteins are made and deployed is a fascinating story involving several of the cell's internal components and a Nobel Prize, but we'll save that for later in this chapter, when you're more familiar with those components. For now, we'll use transporter proteins to slip through the cell membrane into the cytoplasm and continue our tour.

Your first reaction is certainly amazement at everything that's going on in here and the speed at which it's all happening. Cells are busy places, crammed full of vesicles, enzymes, and organelles all fluttering around in a frenzy of activity. Motor proteins can carry cargo around at a speed of one micron per second. Considering that vesicles—which are little cargo containers—can be as small as 50 nm across, that would be like a car that is 15 feet long moving 200 miles per hour. These vesicles are being delivered to and from various organelles—literally "little organs"—and to and from the plasma membrane, in the case of cargo exported from or imported into the cell.

These vesicles that are whizzing past us, powered by motor proteins, travel on tracks made of microtubule proteins, which are components of the cell's skeleton. Yes, the cell has a skeleton, or as we sophisticated scientists call it, a cytoskeleton. There are different types of cytoskeletal proteins with different structures and functions. The protein actin makes long, thin microfilaments that can form bundles and networks that stretch the plasma membrane, shaping and supporting the cell, rather like tent poles. The microtubule cytoskeleton is made up of tubulin, which forms, yes you guessed it, little tubes. Microtubules are wider in diameter than actin microfilaments and, because of their hollow shape, are less flexible. Microtubules run throughout the cell like monorail train tracks. The vesicles that move along the microtubules are like the train cars, and the motor proteins are the locomotives. Microtubules have polarity—not electrical polarity, but a specific head-to-tail orientation. There are two general kinds of microtubule motors, each of which only moves in one direction. Cargo, such as a vesicle, must be switched from one type of motor to the other if it needs to change direction.

Microtubule motors carry relatively huge loads of cargo—vesicles or even whole organelles—essentially by "walking" along microtubules

Although each step a motor takes might only move the cargo about 10 nanometers, it can take up to 100 steps per second—which means cargo can move through the cytosol at breathtaking speeds of one micron per second.

Microtubules have other functions, too, like enabling sperm to swim. The tail of a sperm is filled with a long, thin array of microtubules. Motor proteins cause these microtubules to slide relatively to the others, which makes the tail bend back and forth, and the chase is on.

The most important compartment inside the cell is the nucleus. This is where the chromosomal DNA is located and where DNA replication and transcription—the production of RNA from the DNA genetic code—occur. Since this is a human cell, it contains twenty-three paired chromosomes, which store the DNA that serves as the blueprints for this person's body and the directions for how it functions. At the level of the single cell we are touring, the DNA carries genetic information that will be "expressed" into the structures that make up the cell and permit different functions to occur. This DNA is broken up into individual genes, which, taken together, make up the genome. Most cells contain two copies of each gene. In organisms that reproduce sexually, sperm and egg cells contain one copy of each gene, so that the fertilized egg is the first true cell of the offspring, with the requisite copies of each gene, one from mom and one from dad. This fertilized egg, or zygote, then undergoes numerous rounds of division, and these cells soon begin to differentiate into the many specific organs and tissues found in the fully developed adult.

The nucleus directs the activity of the cell by sending newly synthesized messenger RNA (mRNA) out into the cytosol where it is translated into functional protein by the ribosome. The nucleus has a double membrane, referred to as the nuclear envelope, two bilayers separated by a thin perinuclear space. Because of this double-membrane structure, transport in and out of the nucleus does not occur via vesicles. (A benefit of this is that viruses cannot deliver their genomes into the nucleus directly by fusing with the nuclear envelope.) Instead, there is a large channel that connects the nucleus to the cytoplasm. This nuclear pore complex is made up of about thirty different proteins, each with generally sixteen copies per pore in a radially symmetrical array. There are thousands of these pores in each nucleus. The inner diameter of the nuclear pore complex is about 50 nm, which is ten to twenty times larger than most channels in other cellular membranes. Although the pore is large, transport in and out of the nucleus is tightly regulated. This is how mRNA gets out of the nucleus and how things that regulate DNA replication and RNA transcription get in. Like cargo trucks, a class of proteins called karyopherins—specifically, importins and exportins—bind cargo and mediate the traffic into and out of the nucleus.

For the next stage of our tour, rather than float around aimlessly from one organelle to the next, let's proceed in a way that shows us how these different components are involved in the production, transport, and breakdown of proteins. Specifically, we'll start with one of those integral membrane proteins we saw on our way in.

Our understanding of how these proteins are made and transported to the plasma membrane is largely credited to a scientist named Günter Blobel. In the late 1960s, Blobel trained in the laboratory of George Palade at the Rockefeller University, then called the Rockefeller Institute. Palade would go on to win the Nobel Prize in Physiology or Medicine in 1974 for his seminal work uncovering aspects of the structural organization of the cell. Palade developed techniques for labeling, visualizing, and measuring the synthesis, trafficking, and modification of newly formed proteins. One of Palade's most sensational discoveries had to do with ribosomes, which are little spherical bodies inside the cell that carry out the translation of mRNA into protein. Palade found that while some ribosomes float around freely in the cytosol, others are affixed to a large membranous compartment within the cell called the endoplasmic reticulum (ER). Palade knew that some parts of the ER appeared rough in his images, rather than smooth like the rest of the ER. As it turns out, ribosomes stuck to the ER made it look rough.

The rough ER is where membrane proteins are made, as are proteins destined to be secreted into the extracellular space. Palade had already learned a great deal about the timing and compartmentalization of the so-called secretory pathway, which is both the route and the process by which proteins made in the ribosomes move to the plasma membrane and beyond. What Blobel figured out was that the instructions for this process, the control of how and where translation would take place, was actually encoded by the newly synthesized protein itself.

As noted, all proteins start out on cytosolic ribosomes. However, if the first few amino acids formed contain what Blobel called a signal sequence, the ribosome is then directed to the rough ER. A signal sequence is a short stretch of amino acids that contains a binding site for what is called the signal recognition particle, or SRP, which floats around the cell looking for ribosomes with newly synthesized signal sequences sticking out of them. The signal sequence binds to the SRP, which, like a tugboat, tows the ribosome to the outer face of the ER, where it docks to an SRP receptor.

The ribosome then connects to a large channel in the ER membrane called the translocon. The translocon is like a huge tube through which the newly synthesized proteins exit the ribosome into the ER. The exit site in the ribosome for newly synthesized proteins becomes precisely aligned with the pore of the translocon, and as the amino acid chain is elongated by the ribosome, it is fed through the translocon. If the ribosome is carrying a protein meant for secretion—like a growth factor, a neurotransmitter, or a hormone such as insulin— this cargo is off-loaded into the fluid-filled space inside the ER. Alternatively, if the newly synthesized protein is meant to be an integral membrane protein, it will contain at least one transmembrane domain, a stretch of hydrophobic amino acids that will remain within the lipid bilayer surrounding the endoplasmic reticulum. Somehow, the transmembrane domain will then leave the protected confines of the translocon pore laterally and enter into the lipid membrane surrounding the ER. How this actually occurs is not entirely clear.

Although proteins are synthesized as linear chains of amino acids, they adopt complex three-dimensional shapes that are required for their proper function. This process of protein folding generally occurs with the help of "chaperone" proteins that reside inside the ER and, like parents cradling babies in their arms, protect the sensitive, newly synthesized proteins. If protein folding goes wrong, the result can be a nonfunctional ball of junk, and this can set off a chain reaction in which large numbers of incorrectly formed proteins form even larger balls of junk that can ultimately damage the cell and cause disease. This is an area of intense research, as protein aggregation seems to underlie debilitating human diseases such as Alzheimer's and mad cow disease (which in humans is called Creutzfeldt-Jakob disease).

Once proteins are properly folded, they move on along the secretory pathway, which becomes a bit like an assembly line, adding to and modifying the proteins. Many proteins are altered by enzymes by what are called post-translational modifications. For example, the process of glycosylation adds carbohydrates to proteins, sort of like little antlers made of sugars. The extracellular domains of integral membrane proteins that are headed to the plasma membrane can be heavily glycosylated. Some cells have so many carbohydrates on their surfaces that they are literally sugarcoated with a glycocalyx that serves as a protective barrier.

The next stop along the secretory pathway is the Golgi apparatus, a membrane-bound organelle where particular post-translational protein modifications take place. These include further glycosylation events, as well as the addition of small molecules such as phosphate groups that include a phosphorus atom along with four oxygen atoms. Phosphate groups are very important signals that reflect the activity status of proteins and mediate interactions between proteins. Eventually proteins emerge from the Golgi like college graduates—fully folded, modified, and mature—ready to be moved to the locations where they can do the jobs for which they were made. These proteins exit through the trans Golgi network (TGN), which is the birthplace of countless vesicles that bud off from the Golgi laden with newly synthesized secretory protein cargo. Some vesicles are destined for the plasma membrane to release proteins into the extracellular space, or to transfer integral membrane proteins into the plasma membrane. Other vesicles carry proteins needed by various organelles inside the cell.

Now let's look at inbound traffic. Endocytosis is the process by which cells internalize cargo from the extracellular space. Some of this cargo has to be degraded inside the cell if it cannot be employed for a beneficial purpose or "recycled" back out of the cell. Otherwise, it could build up and lead to toxic effects. This degradation occurs inside the lysosome, an organelle filled with acids and degradative enzymes that break apart proteins. These enzymes are actually formed in the secretory pathway and get into the lysosome via vesicles that bud off of the TGN.

In the Golgi, proteins that are destined for transport to the lysosome are modified by the addition of a carbohydrate called mannose-6-phosphate. Upon exit from the TGN, these proteins encounter the mannose-6-phosphate receptor, which is like the ticket taker on the train to the lysosome. If there is an error in any step of this process, for example a mutation in one of the enzymes involved in the formation or addition of mannose-6-phosphate, lysosomal storage disease can result. Lysosomal storage diseases such as Tay-Sachs disease are debilitating, painful, and even deadly. Fortunately, they are also rare.

In summary, the cell is a busy and complicated place with a vast, complex transportation system connecting organelles and compartments via microtubules. Newly formed integral membrane proteins travel through the secretory pathway from the ER to the Golgi and eventually to the plasma membrane. Later, the same membrane proteins will undergo endocytosis, moving inward to be broken down in lysosomes. Turnover and replacement of proteins is a constant process. Moreover, endocytosis of integral membrane proteins can be stimulated by outside factors, particularly in the case of receptors that respond to extracellular growth factors that tell the cell to divide. However, it can be dangerous if cells are too fruitful and multiply too much.

Activated receptors have to be moved back inside the cell so that the signal to reproduce can be stopped. Otherwise, the cell would replicate nonstop, which would result in cancer. Thus, the receptor that was carried outward through the secretory pathway to the plasma membrane to respond to growth factors outside the cell is now routed back to the lysosome, where it will be degraded—that is, taken apart the way a junked car is disassembled for spare parts and recycling.

Now we've seen that there is a tremendous amount of complex activity going on inside this tiny cell. We've seen proteins being built and torn down and moved into, out from, and within the cell at amazing speeds via an incredibly complex transportation web.

The question you might now be asking is this: Where does the energy come from to power all this activity? All cells are powered by a basic unit of energy, a molecule called adenosine triphosphate, or ATP. These high-energy molecules are produced by the cell's power plants, organelles called mitochondria.

Although it was long known that mitochondria were the site of ATP synthesis, nobody had any idea how this actually occurred until 1961, when Peter Mitchell published a hypothesis in the journal *Nature*. Mitchell had no experimental evidence, and most scientists did not take his ideas very seriously. However, time would vindicate Mitchell; his work earned him the Nobel Prize in Chemistry in 1978.

Mitchell proposed the chemiosmotic hypothesis for ATP synthesis. Osmosis is the movement of water from an area with a low concentration of diluted molecules, such as sugars and salts, into an area with a high concentration of diluted molecules, across a barrier that allows the water but not the diluted molecules to cross. If you put dried fruit, like a prune, into a glass of water, it will take on the liquid and swell up. That is osmosis. The salts and sugars in the prune attract the water. Chemiosmosis is when the diluted molecules, not the water, move across the barrier. This movement of molecules, such as dissolved charged ions, can trigger things like muscle contraction and signal conduction along nerves.

Mitochondria are strange organelles that, like the nucleus, have two bounding membranes instead of one. The space between the two membranes, the periplasm, contains a higher concentration of protons—positively charged hydrogen ions—than the inside of the mitochondrion. Mitchell hypothesized that a series of enzymes resides in the inner mitochondrial membrane to more or less pump protons from the inner space into the periplasm. The energy for this process is provided by the breakdown of carbohydrates, fats, and proteins found in the food we eat. Mitchell predicted that the protons that were pumped into the periplasm would be allowed to flow back into the inner core of the mitochondrion through a channel in the inner mitochondrial membrane. There was method to this seeming back-and-forth madness, moving protons in and out of the periplasm. Mitchell proposed an enzyme, now referred to as ATP synthase, which would form ATP, powered like a waterwheel driven by a flow of protons. Thus, energy that comes into the cell in many forms is elegantly converted into ATP, a single, universal unit of energy that powers all the body's reactions and processes.

Mitochondria have transporters similar to the nuclear pore complex that allow selective transport with the cytosol. However, unlike the nucleus, the periplasm between the inner and outer mitochondrial membranes is the site of a great deal of activity. For example, this is where the protons are transported to power ATP synthase. While the nuclear membrane has pores that transport molecules directly into and out of the nucleus, many substances move through the mitochondria's periplasm in two stages, first across one membrane, and then the other. It's a bit like passing through an airlock, except that certain molecules are meant to remain in the airlock—the periplasm—while others need to pass through the inner membrane to the inside of mitochondria.

This process of mitochondrial transport is very much like the way things move in and out of many bacterial cells. In fact, it is believed that mitochondria actually are the adapted vestiges of a previous stage in evolutionary history where a larger cell internalized bacteria. This endosymbiotic theory is supported by the fact that mitochondria actually contain their own DNA that encodes some of the proteins required for ATP synthesis. As in bacteria, mitochondrial DNA is constructed in one large circular piece, as opposed to the linear chromosomes of nuclear DNA.

Mitochondrial transport is generally very well controlled. However, during the process of apoptosis—programmed cell death—the mitochondrial membrane becomes leaky. This increased permeability is one of the hallmarks of apoptosis, which is critical for the development and survival of multicellular organisms. Damaged or diseased cells must be removed. This prevents viral infections from spreading and stops the progression of cancer. Normal growth and development require apoptosis as well. For example, our hands and feet are webbed like a frog's while we are early embryos. Apoptosis normally removes the cells between fingers and toes, and if this doesn't occur, the condition known as syndactyly results. The pathways that lead to apoptosis, such as a sudden increase in mitochondrial permeability, are very tightly regulated and critical for survival. They'll be discussed in detail in a later chapter.

So now you've had a pretty good look at the cell, its internal structures, how proteins are made and unmade, how stuff gets in and out and moved around, and how cells communicate with the rest of the body. But we've hardly touched on what makes all this happen—the central store of code that tells the cell what it is and what it's supposed to do and how these orders are communicated. And so, on to the next part of this book . . .

PART TWO

THE GENETIC CODE—
HOW IT WORKS, WHAT
IT MEANS, AND HOW
HUMANS CRACKED IT

CHAPTER 3

The Central Dogma of Molecular Biology

W HAT DO WE REALLY MEAN BY "MOLECULAR BIOLOGY"? Luckily for us, that is a question with one of the most definite answers in all of science.

"Central dogma" might sound like a phrase from an Orwell novel, but it actually refers to the basic principle of molecular biology. This whole field of biological study can be summed up simply in one concise phrase:

DNA becomes RNA becomes protein.

Of course, the devil is in the details, and there are numerous addendums to this modest abstraction. Nonetheless, these five little words are the first thing you need to learn about molecular biology and the foundation of its study.

DNA, of course, is the acronym for deoxyribonucleic acid. Let's parse that out:

- "nucleic" means it's (generally) found in the cell's nucleus
- "ribo" means the structure of DNA is based on a five-carbon ribose sugar
- "deoxy" means the ribose sugar is missing an oxygen atom
- an acid is any molecule that, at neutral pH, will lose a proton, which is a subatomic particle with a positive charge

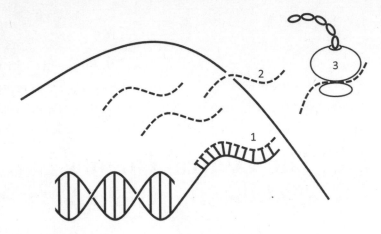

(1) DNA (left) in the nucleus is transcribed into messenger RNA (mRNA; right). (2) The mRNA is exported from the nucleus to a ribosome (3), where it is translated into a polypeptide protein.

Thus, in some ways, DNA is named for its deficiencies. RNA stands for ribonucleic acid—a similar molecule, but one that hasn't lost that oxygen atom. Nonetheless, DNA and RNA are quite distinct from each other in structure and function. The structure of DNA is a double helix, while RNA forms a single helix. RNA is also a much more sensitive molecule, quickly degraded in the wrong conditions. DNA, on the other hand, can stay structurally intact for very long periods of time. In 2013, an international team of researchers was actually able to isolate and sequence small fragments of DNA from a sample of a 300,000-year-old fossilized bone from a cave bear found in Spain. Not exactly *Jurassic Park*, but astounding nonetheless.

What does the central dogma really tell us? What exactly is DNA and how does it become RNA? What is protein, other than something bodybuilders drink in powder-based beverages?

Think of a restaurant. The table you are eating at is the nucleus, and the menu is the genome—all of the DNA in the cell. The genome is not an incoherent jumble of DNA; it is organized into individual genes. Each gene is like a dish on the menu, a Caesar salad, a burger, lentil soup—and just as the burgers at two different restaurants are never exactly the same, there are usually subtle differences in the versions of genes from one individual to the next. The exception to this would be clonal restaurants like McDonald's, identical siblings without any real genetic differences to produce distinguishing characteristics.

At most sit-down restaurants, how do the cooks know what you want to eat? You don't take your menu into the kitchen and tell them what you want, and the cooks don't come to your table to ask. A server takes your order, writing it down on a slip of paper. This is just like the mechanism, the reaction that synthesizes RNA, and the RNA is the slip of paper on which your order is written. A particular gene is transcribed into RNA and then this gets expressed as a protein, the meal that actually comes out of the kitchen. Interestingly, most items on the menu at the Central Dogma restaurant are not precisely fixed. You can add cheese and bacon to your hamburger if you want, but more on that in a bit.

The RNA transcript exits the nucleus and goes to the ribosome, which is the kitchen—the place in the cell where the RNA is translated into protein. The ribosome is made of two pieces that look a bit like one mushroom cap on top of another slightly smaller one. The RNA molecule slides in between them. The protein is made inside and snakes out through a small hole in the larger mushroom cap. Like a kitchen staffed with experienced line cooks, the ribosome ensures that the individual subunits of a protein, the amino acids, are correctly arranged according to the sequence of the RNA. This involves specific selection of components as well as putting the proper pieces in the appropriate arrangement. You wouldn't want oatmeal inside your omelet (the wrong components) or the ham and cheese on top of the omelet, rather than inside it (correct components, but in the wrong order).

In the first step of ordering up a protein, each appropriate piece of the DNA code is transcribed into a corresponding RNA molecule, each gene thereby becoming a transcript. Both DNA and RNA are polymers—linked chains of small molecules called nucleotides. There are four different basic types of nucleotides. In DNA, they are referred to as thymine, adenine, guanine, and cytosine. In the case of RNA, the nucleotide uracil is used in place of thymine, although the two are extremely similar. All these nucleotides are small carbon-based ringlike molecules that differ from one another by just a few atoms in the portion of the nucleotide known as the nitrogenous base—one nitrogen or oxygen molecule added or subtracted is all the difference between them.

Thanks to James Watson and Francis Crick, and Maurice Wilkins and Rosalind Franklin (who are often left out), we now know that DNA is shaped in a double helix, a spiral ladder formed by two opposing strands. Each rung of the ladder is a bond formed between the nitrogenous base portions of two complementary nucleotides. Adenine (A) binds with thymine (T), and guanine (G) binds with cytosine (C). No other combinations are permitted. This is referred to as "Watson and Crick base pairing." Wherever you have an A on one strand, it is bound across to a T on the other, and wherever you see a G, it is linked to a C. This is the general structure of DNA then, two long polymers, chains of nucleotides, which are bound together through base pairing across the gap between them. Imagine two strings of magnetized beads sticking together. Importantly, though, the bonds that serve to hold together the two strands—the ladder's rungs—are not as strong as those that link adjacent nucleotides within a single strand. Thus, the double helix can separate, leaving two single-strand chains of nucleotides, which is necessary for DNA replication and transcription.

The reason paired DNA strands separate from each other is so one can serve as a template from which a complementary RNA strand can be polymerized. The specific sequence of the gene being transcribed is copied as the RNA polymer is created. The transcription reaction, which involves a group of enzymes collectively called the RNA polymerase, produces the RNA polymer from individual nucleotide units.

After the two opposing strands of the DNA double helix have separated, the RNA polymerase moves along the DNA and reads the specific order of nucleotides along one strand, called the template. (The other DNA strand is called the coding strand, as it is essentially identical to the RNA that will "code" for the protein.) At each position on the template, the enzyme adds the complementary RNA version of the DNA nucleotide it encounters. For example, a DNA template sequence would produce a complementary RNA sequence like this:

DNA: T G A C A C T G

RNA: A C U G U G A C

In this way, the new RNA strand is identical to the segment of the DNA coding strand that was formerly bound to the template. The newly formed RNA now carries the genetic code that will determine the sequence of the protein that will be formed. Wherever there is a G in the template, RNA polymerase adds a C to its own polymer, making a match, and so on. If there are any mismatches, if the RNA polymerase makes an error, the absence of a correct base pairing is detected and corrected.

Now that the RNA is transcribed, one more step is needed before it makes its journey from the nucleus to the ribosome to be translated into protein.

The genome is made up of chromosomes. Humans have 46 chromosomes, half of which come from the mother and half from the father. Each chromosome is a double-helix polymer of millions of nucleotides paired in a linear sequence. This sequence is segmented into individual genes, which are transcribed into specific RNA molecules to be translated into particular proteins. This is the basis for the "one gene, one protein" hypothesis, which, although basically correct, does not adequately describe the potential variability in the system. In actuality, many different versions of a protein can be created from the same gene. This variation in protein sequence and structure is mostly controlled at the RNA stage immediately following transcription.

So how does this happen? How can one person get his burger without onions while another can get hers with extra pickles?

A gene is a segment of DNA that produces—via transcription and then translation—a single protein. But how can a single gene make a number of different versions of a protein? This occurs because genes are actually made up of different modules that can be combined in different ways.

Genes are made up of two different kinds of DNA code. The segments of DNA that actually encode the protein being created, the expressed sequences, are called exons. The intervening sequences, which separate adjacent exons, are called introns. While genes (DNA) and newly synthesized transcripts (RNA) contain both exons and introns, the final, fully mature messenger RNA (mRNA) that feeds into the ribosome during protein synthesis contains only exons. The introns are removed by a reaction known as splicing.

Imagine that the genetic DNA code in the chromosome is an old-fashioned audio tape recording of a live concert, and in between each song the band takes some time to retune or decide what to play next. You could cut out the bits of tape between the songs, splice the tape back together, and end up with a recording that only has the songs with nothing separating them. This is analogous to how the introns go away in the process of making mRNA. But beyond that, during gene expression, sometimes some of the exons are left out as well. This is called alternative splicing, and it happens all the time. Different blocks of introns and exons can be spliced out. Sometimes certain exons are spliced out along with the flanking introns, and sometimes two different exons are mutually exclusive so that only one can be expressed. It's like cutting out some of the songs you don't like from the concert tape, or selecting the version you like best if the band played the same song multiple times. Alternative splicing greatly increases the diversity of the genome. The same gene can be expressed several different ways, making a variety of specific proteins. As an extreme example, it has been calculated that the DSCAM gene from the *Drosophila* fruit fly can be spliced in over 38,000 different ways. The different proteins encoded by splice variants can display alternative functions, localizations, expression patterns, and stabilities. Some genes are expressed as particular splice variants in some tissue or cell types relative to others, while other specific splice variants are strongly linked to certain diseases such as cancer.

Once splicing has occurred, the fully mature mRNA can now exit the nucleus in search of a ribosome, where translation of the mRNA molecule into a protein takes place. A protein is basically a long chain of amino acids that are linked together by peptide bonds. We refer to the protein that gets spat out of the ribosome as a polypeptide not only because of its structure, but because a polypeptide is not exactly a protein per se. Calling it a protein would suggest that the polypeptide has attained its final functional structure. However, proteins generally do not look like simple linear chains of amino acids. Rather, the polypeptide chains fold into complex shapes from which the protein's function follows. When you order a cheeseburger, it doesn't come to you with the meat, cheese, onions, pickles, and buns splayed out on a plate; the ingredients have to be properly assembled and stacked in a certain order. Furthermore, many proteins are generated by the combination of several different polypeptides. These are called multimers and can be groups of the same polypeptides—homomultimers—or groups of different polypeptides—heteromultimers—that work together. This is like combining separate components into a single dish, like three small "slider" cheeseburgers on the same plate.

Finally, let's have a look at what actually happens inside the ribosome once the mRNA arrives there.

The coding sequence of a gene that gets transcribed into an mRNA molecule can be thousands of nucleotides long. But no matter how long the coding sequence is, it will always be in multiples of three, because it takes three nucleotides to denote a single amino acid. These three-nucleotide units of genetic code are called codons. There is a specific "start" codon that tells the ribosome to begin a new polypeptide. The start codon sequence is AUG—adenine-uracil-guanine. (AUG in the mRNA molecule, you will remember, corresponds to ATG in the coding strand of the gene.) Likewise, there are three "stop" codons. In addition to these, there are specific codons for all of the 20 amino acids commonly found in proteins. There are 64 different possible codons, as there are four possible nucleotides at each of the three positions ($4 \times 4 \times 4 = 64$). As there are only 20 different amino acids, each is generally represented by more than one codon.

Inside the ribosome, the mRNA molecule advances one codon at a time. As each codon is recognized, it is bound via Watson-Crick base pairing to a small complementary piece of RNA that is bound in turn to a single amino acid. This transfer RNA (tRNA) binds to the codon with its own "anti-codon" and holds the appropriate corresponding amino acid in place just long enough to allow a peptide bond to be formed between the growing polypeptide chain and that amino acid. The ribosome connects the pieces together in a specific order based upon a series of instructions.

Sometimes mistakes occur in this process. A misread codon leads to an error in protein sequence if the wrong amino acid is selected. That said, a single incorrect polypeptide can't do much damage. The big problems arise when the original DNA sequence is mutated, either having been passed down from a parent or arising during development when new cells are being rapidly produced and the genome with the mutation is duplicated repeatedly. If a mutation incorrectly creates a "stop" codon toward the beginning of a gene rather than at the end, a viable, functional protein will not be produced.

One safeguard against mutations is that you do have a maternal and paternal copy of each gene, so if one is mutated, the other should be able to compensate. A "dominant" mutation is one that has a negative effect in spite of being paired with a normal gene from the other parent. A "recessive" mutation is one that isn't evident, because the presence of the normal gene from the other parent is sufficient. A recessive mutation only causes a problem in the individual when a mutated version of the gene is inherited from both parents. Whether a mutation is dominant or recessive depends both upon the gene (whether you need all the protein produced from two functional copies) and the specific change in question (whether it actually inhibits function or creates some new damage or malfunction).

There are multiple ways in which we are protected every day from the damage that mutations and other errors in gene expression can generate. Of course, there are still many ways that problems at the level of molecular biology can indeed lead to debilitating disease. However, before we get too deep into genetics, let's go back to the beginning. How did we ever figure out that DNA is the molecule that carries genetic information in the first place?

CHAPTER 4

Solving the Mystery of Life: The Road to the Double Helix

T ODAY WE ALL UNDERSTAND THAT AN INDIVIDUAL'S GENOME IS encoded by DNA. We know that DNA is concentrated in the nucleus, the largest subcellular compartment or organelle. And we know that within the nucleus, DNA is arranged in a highly organized fashion. Chromosomal DNA winds around proteins called histones, like thread on a spool. How tightly the DNA is wound around the histones helps determine the ease with which the genes in the DNA can be expressed. The looser the loops of DNA, the easier it is for the RNA polymerase to gain access to the gene sequences and mediate transcription.

A hundred years ago, scientists hadn't figured all this out. They knew that whatever carried the genetic information of a higher organism was most likely located within the nucleus, but the prevailing opinion was that the proteins, not the nucleic acids, were responsible. Also, they believed that genetic processes, the ability to make changes that can be inherited by a new generation via reproduction, were limited to higher organisms, such as humans, with nuclei, histones, and chromosomes. This was based on the observation that bacteria don't even have nuclei, nor is their DNA arranged in chromosomes wound around histones. That paradigm was shattered

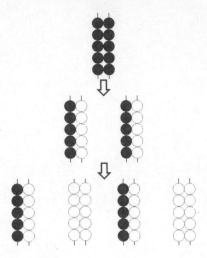

"The most beautiful experiment"—the semi-conservative
model of DNA replication as demonstrated by Meselson
and Stahl.

in 1928 when Frederick Griffith published a paper entitled "The
Significance of Pneumococcal Types" in the *Journal of Hygiene*. This
seemingly incremental event led to nothing short of a revolution and
the creation of an entirely new field of science known as molecular
genetics.

Griffith was a medical officer in the British Ministry of Health
and was interested in understanding how bacteria caused pneumonia. He was studying two types of pneumococcus bacteria. One was
referred to as smooth (S) and the other rough (R), due to the way
colonies of each type of bacteria look when grown in a lab. The difference in structure and appearance arises from a carbohydrate coating that the S has but the R lacks. This coating prevents the host's
immune system from attacking the bacteria, so the S is virulent and
can cause disease, while the R cannot, as it is visible to the immune
system and can be identified and targeted for destruction. So if you
give a lab mouse the S strain, it will die, while inoculation with the R
strain will not make it ill at all.

What Griffith did was to combine the R, the non-virulent bacteria, with a sample of the S strain that he had heated up to the point where it died and was therefore harmless. Live S was lethal when injected into mice, while live R was not. However, if you mixed live R with the S that had been killed by heat, somehow something associated with a dead sample of S would "transform" the R into S, and the mice would die. In Griffith's words, "The inoculation . . . of mice of an attenuated R strain . . . together with a large dose of [S] . . . killed by heating to 60°C, has resulted in the formation of a virulent S pneumococcus . . ." Furthermore, live S could be cultured from the blood of the infected mice. Thus, the transformation of the non-virulent R strain into the deadly S strain was heritable: the progeny of those live bacteria were additional generations of virulent S-strain bacteria. This was an astounding finding. Something that was present in the sample of dead bacteria was able to convert the living non-virulent R bacteria into the S form. This meant that bacteria did indeed undergo genetic changes. However, the remaining key question—what part of the heat-killed S strain had caused the transformation of the R—wouldn't be solved for another fifteen years, when scientists on the other side of the Atlantic would figure out what Griffith called the "chemical nature of the transforming principle" that changed the R into S.

In February 1944, three researchers at the Rockefeller Institute in New York City published one of the most elegant and significant series of observations in the history of biology. While several important steps had recently been taken in this area, such as the replication of Griffith's observations in liquid growth media in a test tube, rather than in a living, soon-to-be-dead mouse, Oswald Avery, Colin MacLeod, and Maclyn McCarty proved that DNA was the material of genetics—Griffith's transforming principle.

In order to get their experiment to work, Avery, MacLeod, and McCarty had to take great pains to eliminate factors that would prevent or mask their astounding result. They determined steps to reduce the presence of residual R cells in their cultures, so that the S cells, if they were produced, would be detectable. In other words, they sought to decrease the size of the haystack to more make the needle easier to find. They did this by adding antibodies that would bind the R cells to their culture, causing them to stick together and form large aggregates that would fall out of solution and collect at the bottom of the culture. Antibodies are proteins created by our immune system to mark and identify foreign invading pathogens. They bind very tightly and specifically to their cognate antigens, in this case the surface of R cells. This would leave any newly formed S cells floating and growing so that they could be collected.

The scientists also determined conditions required to inactivate an enzyme from the S cell preparation that would otherwise degrade the factor responsible for transformation. An enzyme is a protein that performs some sort of work. In this case, there was some enzyme that was preventing transformation, which they needed to get rid of. If they could figure out exactly what the enzyme was degrading, they would have their answer as to what mediated the R-to-S transformation.

Avery, MacLeod, and McCarty made a lysate of the heat-killed S strain, that is, a liquefied mixture of all the cellular components, sort of a bacterial smoothie. Then, through a combination of fractionation (a crude purification of different types of molecules) and enzyme treatment, they determined that DNA was the transforming principle. They showed that carbohydrates, proteins, and RNA were definitely not capable of transforming R cells into S cells on their own. When they isolated a fraction rich in DNA, using techniques still employed today, they found that the DNA would transform the bacteria. However, if they treated this fraction with an enzyme preparation that degraded DNA, transformation no longer occurred. Thus, they definitively proved that DNA was the transforming principle, the material that allowed transformation to occur. This was such a shocking result that for years people attempted alternative means to test it. One of the most famous examples of a follow-up to Avery, MacLeod, and McCarty was published in 1952 by Alfred Hershey and Martha Chase.

As mentioned earlier, a phage is essentially a very simple virus that infects bacteria. Hershey and Chase chose a phage called T2 that infects *E. coli* bacteria, and they labeled all of its DNA with radioactive phosphorus. Phosphorus is not normally found in any of the amino acids that make up proteins, but it is present in DNA. In a separate sample, they labeled all of the phage's protein with radioactive sulfur. Sulfur is found in some amino acids, but not in DNA. Hershey and Chase then infected *E. coli* cells with the two differently labeled phage. Voila! The phosphorus-labeled phage DNA became intimately incorporated into the host bacteria, while the sulfur-labeled phage protein could easily be removed from the infected cells. Furthermore, new phage particles produced from infected bacteria contained a great deal of radioactively labeled DNA, but almost no labeled protein. Hershey and Chase concluded, "This protein probably has no function in the growth of intracellular phage. The DNA has some function." However, they stopped short of explicitly stating that DNA was the genetic material—Griffith's transforming principle. The longtime assumption that protein was the molecular basis of genetics was still too deeply entrenched.

Only a year later, though, in 1953, American James Watson and Briton Francis Crick issued a groundbreaking and astounding publication that soon led to a full understanding of the structure, function, and behavior of DNA.

A great deal has been written about Watson and Crick and their seminal discovery. However, as some say about Lee Harvey Oswald, they did not act alone. Although they were certainly informed by competing alternative models for the structure of DNA—such as Linus Pauling's triple helix, which incorrectly placed the nucleotide bases on the outside of the structure of DNA—the main concern from a historical and ethical perspective is the work of their close colleagues. Watson and Crick did acknowledge Maurice Wilkins and Rosalind Franklin in their paper "Molecular Structure of Nucleic Acids: A Structure for Deoxyribose Nucleic Acid." However, Wilkins did not share in the authorship of the paper or in much of the subsequent notoriety, although Wilkins was awarded a portion of the 1962 Nobel Prize. (Franklin died in 1958, so she was not eligible to share the Nobel Prize with Watson, Crick, and Wilkins. There is no way to know whether her contributions would have been rewarded had she lived.)

One fascinating fact about the work of Watson and Crick is that their paper doesn't actually contain any real data. It is only one page long and reads more like the solution to a question on an exam than the results of experimentation. They raise a problem—what is the structure of DNA?—discuss what is known, what solutions have been previously proposed, and why those alternatives are "unsatisfactory" or "ill-defined." Then they move the field—and the history of the human race—a giant step forward logically, basically through thinking out loud. It is a bit like having someone explain how to solve a puzzle or how a magic trick is performed. Once you understand what they are saying, it seems as if the answer was obvious all along and no other possible solution could have ever existed.

Watson and Crick covered a lot of ground in a single page and one line drawing. They proposed that the structure of DNA is a double helix with the phosphate backbones that provide the linkages between the successive nucleotides in each polymer strand on the outside, and the bases, the business ends of the A, T, C, and G molecules, bound together toward the inside of the structure. They suggested that the opposing strands were held together by the base pairing at each rung of the ladder and correctly determined that the two strands of DNA in the double helix pointed in opposite directions—antiparallel. Finally, in an infuriatingly oblique flourish, they stated, "It has not escaped our notice that the specific pairing we have postulated immediately suggests a possible copying mechanism for the genetic material." However, the reader would have to wait until later for the "full details . . . [to] be published elsewhere." One month later, in their follow-up paper, Watson and Crick indeed showed how A and T, and C and G, bound together, only in those combinations, in so-called "Watson and Crick base pairing." Although this demonstrated the molecular basis for the base pairing glue that held together the strands of the double helix, it would actually take another five years for the mechanism of DNA replication to be worked out, and Watson and Crick themselves would not be directly involved.

Matthew Meselson and Frank Stahl are credited with what is widely referred to as "the most beautiful experiment" in the history of biology. This phrase, as quoted in Horace Freeland Judson's book *The Eighth Day of Creation*, was coined by John Cairns, former director of the Cold Spring Harbor Laboratory, a post subsequently held by James Watson. Cairns refers more to the elegance of Meselson and Stahl's brilliance in describing the process by which DNA is copied, rather than any aesthetically pleasing elements of their data. Every organism begins as a single cell, which divides over and over again. In each division, the DNA must be copied exactly and rapidly. Watson and Crick's obscure suggestion needed to be expanded and explained: How could one double helix actually be duplicated into two? How does DNA replication occur?

Three models were proposed.

In the *conservative* model, an entire new double helix identical to the first would be created. The original double helix would be left intact (conserved) following production of the newly copied, completely new DNA molecule.

In the *semi-conservative* model, the double helix would peel apart into the two separate DNA strands, breaking the base pairings. A new partner would then be synthesized for each strand, so that two new "daughter" double helixes would be generated, each composed of one old strand and one new one.

In the final possibility, the *dispersive* model, the parent DNA molecule would be broken up into smaller pieces that would then be stitched back together in such a way that two new whole double helixes would be created, each containing some old DNA and some new.

The experiment that Meselson and Stahl conducted was so simple and ingenious that there was no chance of misinterpretation. All they did was grow some *E. coli* cells in the presence of a stable isotope of nitrogen, an atom containing a different number of neutrons than usual, which thus had a higher mass than normal nitrogen. This "heavy" nitrogen would be incorporated into all the DNA in all the *E. coli* cells and could be easily distinguished from normal, or "light," nitrogen. They then placed the *E. coli* into growth media containing only "light" nitrogen, and allowed the bacteria to divide and replicate their DNA. They isolated DNA from the bacteria at different intervals of time after the switch from heavy to light nitrogen, following subsequent rounds of cell duplication.

If the conservative model were correct, the double helix made of two heavy strands would remain intact indefinitely as new double helixes made solely of DNA containing light nitrogen were copied from it. If the dispersive model were true, the weight of the DNA strands would decrease progressively over time as smaller and smaller percentages of heavy DNA were distributed among the new daughter strands and nucleotides containing light nitrogen were added.

Neither of those outcomes happened. Rather, after one round of cell division, all of the DNA isolated from the bacteria contained 50 percent heavy nitrogen and 50 percent light nitrogen.

After a second round of DNA replication, Meselson and Stahl found that now only 25 percent of the DNA was composed of half heavy nitrogen and half light. The other 75 percent contained only the light nitrogen. In subsequent rounds of division, the amount of DNA containing only light nitrogen kept growing, while the DNA with both heavy and light nitrogen stayed at around the same total amount but in increasingly smaller proportion, as the total amount of DNA grew, for as long as they ran the experiment.

To visualize this, imagine that you have a necklace made of two strings of black beads, representing the DNA strands with heavy nitrogen in the nucleotides. (See the illustration at the beginning of this chapter.) Loose white beads, the only ones available for making new necklaces, represent the normal nucleotides with light nitrogen. If the conservative model had been correct, after one round of replication you would still have had your original necklace with two strings of all black beads, and a new necklace with only white beads. Instead, Meselson and Stahl found that after one round of replication they had two necklaces, each with one string of black beads and a new string of white beads. If you allowed one more round of replication, you would still have two necklaces with one string of black beads and one of white, and also two necklaces with all white beads. Another round and you would have six with all white beads but still only two with one strand of black beads paired with one of white.

This meant that the semi-conservative model was correct. During the first round of DNA duplication, the two strands of the heavy-labeled double helix split apart, unzipped, as it were. Then a new complementary strand made completely from light nitrogen was added to each. In all subsequent divisions, those heavy strands would become paired with new light strands, as only nucleotides made with light nitrogen were available. Similarly, the light strands, when separated from their heavy or light strands, would be joined together with another newly synthesized strand of DNA containing only light nitrogen.

The thirty years between Griffith's demonstration of transformation and Meselson and Stahl's experiment represented a tidal wave of new information completely reshaping our understanding of the genetic basis of life. Many other scientists also made important contributions.

Furthermore, we would soon learn that the central dogma—DNA to RNA to protein—was only the beginning; many other wrinkles in the world of molecular biology would soon become apparent, leading to new ways of understanding genetic diversity and disease.

CHAPTER 5

Epigenetics:
Beyond the Central Dogma

◉

A S WE'VE DISCUSSED, THE CENTRAL DOGMA OF MOLECULAR biology is that DNA becomes RNA, which becomes protein. Or more precisely, genes, which are made of DNA, are transcribed into RNA, which is then translated into protein. DNA is the source code (the blueprints), and the protein is the actual structural and functional machinery of the cell (the building), while the RNA functions as an intermediary (a messenger), the foreman instructing the workers what the architect wants, generally required only for the conversion of the DNA genes into proteins. However, there are other ways of regulating gene expression outside of transcription and translation.

Cells are constantly exposed to altering environmental conditions. When an antibody secreting B lymphocyte is stimulated by the presence of an infectious microbe, its entire physiology must globally shift as rapidly as possible. Similarly, when hormones acutely activate specific target cells to alter particular functions, the expression levels of different genes must be regulated in a very flexible and yet controlled manner. The ability to respond to changing conditions can mean the difference between life and death, and while some of these responses depend on what are called post-translational modifications

Acetylation of histone proteins causes unwinding
of DNA, increasing access for RNA polymerase and
promoting transcription. This mechanism of regulating
gene expression is critical for rapid alterations in cellular
physiology and has been implicated in numerous diseases,
such as cancer.

(the alteration of protein structure and function), a great deal of biological plasticity derives from which genes are expressed by specific
cells at certain amounts.

As we mentioned previously, DNA winds around proteins called
histones like spools of thread. This is an efficient way of packaging
the approximately three billion nucleotides present in the human
genome. The extent to which DNA is wound around histones also
has important implications for gene expression, as it is very difficult
to transcribe a gene that is tightly wound around a histone. So the
ability of a gene to be expressed can be regulated by the extent to
which it is, or isn't, associated with histones.

There is a class of enzymes referred to as histone acetylases, and,
conversely, one called histone deacetylases. These work in concert
to add and remove very small acetyl groups from histone proteins,
respectively. An acetyl group is simply two carbon atoms, an oxygen
atom, and three hydrogens, but the presence of acetyl groups on histone proteins is enough to regulate the interaction with DNA, by
decreasing how tightly the DNA is able to be wound. Having the
DNA no longer tightly stuck to the histones makes it more acces-
sible to the RNA polymerase and easier to transcribe. Thus, histone

acetylation is a mechanism by which gene expression can increase, and by targeting particular sites on specific histones for acetylation, the expression of genes found in the corresponding regions of DNA that bind those areas can be regulated.

Similar to the regulation of gene expression by adding small acetyl groups to histone proteins, DNA itself can be modified with even smaller methyl groups. DNA methylation refers to modification by addition of just a single carbon and three hydrogens, a methyl group. However, methylated genes are much less likely to be expressed; while histone acetylation promotes transcription, DNA methylation inhibits it. In fact, methylation is responsible for the phenomenon known as "gene silencing," where expression of particular genes ceases due to a complete absence of transcription.

So, by regulating gene expression in this way, mechanisms such as histone acetylation and DNA methylation exist as alternative addendums to the central dogma, not predicted by a conventional understanding. What is ever more surprising is that these alterations to DNA and histones, and the regulation of gene expression they confer, can be retained when the genome is duplicated during cell proliferation. This observation that the mechanisms controlling gene expression are heritable has amazing ramifications. Not only will one cell pass on these regulatory controls to a daughter cell following cell division, but, during reproduction, the next generation of organism can retain these chromosomal modifications. Thus, as these mechanisms that can alter gene expression can be inherited, they are referred to as "epigenetic." This is because although histone acetylation and DNA methylation do not function through conventional inheritance, it is not a question of inheriting one particular gene sequence or another; the regulation of gene expression imposed by these mechanisms can be passed down to future generations, as these modifications are replicated along with the associated chromosomal DNA. Some absolutely astounding effects of epigenetic regulation have been proposed, including that trauma experienced by an individual can lead to changes in gene expression that can then be passed down to his or her offspring. In fact, it has even been suggested that descendants of Holocaust survivors display specific epigenetic patterns as a result of the trauma experienced by their forebears.

During sexual reproduction, this can be seen in the phenomenon of "imprinting," where the particular expression patterns seen in the parent are carried through to the next generation. Similarly, the expression patterns of a particular cell can be maintained once that cell has duplicated. This is because although all cells in an individual contain the same genome, what makes a nerve cell a nerve cell and a blood cell a blood cell is which genes are expressed, as well as at what levels, and which specific splice variants. Thus, epigenetic regulation of gene expression also has significant implications for diseases such as cancer.

There is a class of genes called tumor suppressors, which ensure that improper over-proliferation does not occur. If these were silenced by epigenetic means, the result would be uncontrolled cellular duplication, in essence, cancer. It is beginning to look like the role of epigenetics in diseases like cancer could be extremely significant. In fact, understanding the role of epigenetics in diseases, including cancer, is a major area of current inquiry. In particular, researchers are looking for specific DNA methylation signatures that could be used to identify the most dangerous cancers, or those that might be most resistant or susceptible to a particular treatment.

These different emerging exceptions to the conventional paradigms of genetics and molecular biology are altering our understanding of cell biology; however, what about life without cells?

There is a famous line from the movie *Jaws*: Matt Hooper, a marine biologist played by Richard Dreyfuss, is describing great white sharks to Roy Scheider's character Martin Brody, a cop. Hooper says, "All this machine does is swim and eat and make little sharks, and that's all." That is similar to viruses, only without the swimming and eating. In fact, according to most definitions, viruses aren't even technically "alive." They are simply a collection of molecules (basically just proteins, DNA or RNA, and lipids) configured in such a way as to induce a suitable host cell to allow them to be reproduced. That notwithstanding, in many ways viruses represent the apotheosis of the "selfish gene" hypothesis laid out by evolutionary biologist Richard Dawkins.

To paraphrase Dawkins, what this basically means is that, in an evolutionary sense, the "meaning of life" is having genes passed on to a new generation. Thus, it is the propagation of the genetic material inside an organism, the selfish genes in question, that drive all evolution. Why else would a male praying mantis offer himself up as a post-conjugal meal to his mate, except to improve the odds his offspring will start off with a well-fed mother?

Viruses, however, reproduce asexually; offspring are essentially clones of the parent. The virus can truly only exist to reproduce its own specific genetic material, and yet viruses require a host cell in order to accomplish this. Although the mechanisms that have evolved to facilitate viral replication are truly astounding, as with all adaptations, a great deal can be learned about the present more-advanced versions of things by looking backwards at what came before. In this case, that means things that are not really recognizable as living organisms at all, even for argument's sake, things that still persist inside of our own cells that bridge the world between self and nonself, as the immunologists say.

Most scientists believe that the first virus-like things that evolved are what we call transposons, short for transposable elements. Transposons are basically stretches of DNA in our genome, and those of most organisms, that do not like to stay put. Rather, they hop around, creating novel genetic combinations. A scientist named Barbara McClintock is responsible for much of our understanding of transposons, for which she was awarded a Nobel Prize in 1983. In fact, she is the only woman to ever win an unshared Nobel Prize in Physiology or Medicine. Among her discoveries was a great deal of work on how transposons in maize could regulate the characteristics of corn kernels in ways that could not be explained by the conventional understanding of genetics and molecular biology.

Special enzymes called transposases, which are actually encoded within the transposon sequence itself, work to remove a transposon from one place in the genome and then insert it into a different distinct location. Furthermore, other cellular factors such as enzymes involved in DNA replication or gene expression play a role in transposition. Some transposons are duplicated during the process, and these so-called "replicative transposons" really can amount to viruses that are not able to leave the host. This raises the question, at the most fundamental level, of what is the real difference between a virus and a transposon that mediates replication and movement of the transposase gene to other sites in the genome, other than that the virus has found a way to get out of the host. Everything else that makes a virus a virus can generally be viewed as existing to promote the ability of dissemination of genetic material.

Generally speaking, viruses remain simply a core of genetic material surrounded by a protein coat, called the capsid. Viruses can contain either DNA or RNA genomes. RNA viruses include the retrovirus family, so named because they contain an enzyme called reverse transcriptase that is essential for their life cycle. Reverse transcriptase, or RT for short, does exactly what the name suggests, it converts RNA into DNA. This remarkable addendum to the central dogma of molecular biology essentially acts as a reverse arrow between DNA and RNA, or gene and transcript. RT functions after a retrovirus has entered the host cell, making a DNA copy of the viral RNA. This allows the virus to insert its genome into the DNA of the host cell so that the infection is permanent—at least as long as the infected cell survives and is transcriptionally active. HIV, the virus that causes AIDS, is an example of a retrovirus, and it is this integration of the viral DNA produced via reverse transcription into the host genome that can makes diseases like AIDS so difficult to permanently cure.

RT is also employed in research and biotechnology. Having a catalog of all the RNAs being expressed in specific developmental or disease state can give a very useful molecular snapshot, and create a resource for analysis and experimentation. As was stated earlier, DNA is much more resistant to degradation than RNA. Thus, scientists often want a way to make a DNA copy from a specific RNA or group of RNAs so that it can be easily stored and employed in repeated studies. All the researcher has to do is biochemically purify RNA from a sample and then add RT to create what is known as a "cDNA library." The "c" stands for complementary, as what is generated is DNA that is an exact base-for-base match with the expressed RNAs. Furthermore, if mature mRNA is collected and used to create the cDNA, no introns will be present, as they will all have been already spliced out. This greatly reduces the amount of material relative to genomic DNA and simplifies subsequent analyses.

RT is an example of an enzyme carried into the host cell by the virus; reverse transcription does not normally occur in our cells. While other enzymes required for the viral life cycle can also be found within the capsid, and are immediately available to function as soon as the virus enters the host cell, other proteins essential to the viral life cycle are encoded by the viral DNA or RNA, and only expressed after a cell is infected. However, viruses also generally rely upon proteins present in the host cell, similar to the way transposons can require DNA replication and repair enzymes. As we will learn in the next chapter, reverse transcription is not the only surprising exception to what seem to be fixed rules in molecular biology and genetics.

CHAPTER 6

The Monk's Garden: Mendel's Law of Independent Assortment and How It Can Be Broken

◉

B ORN IN 1822 IN WHAT IS NOW THE CZECH REPUBLIC, Gregor Mendel was a monk whose groundbreaking work— literally groundbreaking, as it was done in the garden of his abbey—resulted in his reputation as the father of modern genetics. Mendel's discoveries are truly remarkable in that neither he nor any other scientist of his time had any way of observing or describing the actual molecular processes underlying fertilization or chromosomal assortment. As you have already read, the mechanism of DNA replication wasn't even correctly described until after Watson and Crick's groundbreaking publication in 1953 regarding the structure of the double helix, almost a century after Mendel conducted his research on plant hybrids.

Educated in science and interested in heredity, Mendel studied, among other things, pea plants. He noticed that there were different traits in pea plants and that under the experimental conditions he determined, seeds that came from one type of plant would usually grow to exhibit these traits in a predictable fashion. The traits he

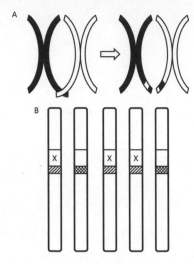

A. Crossing over of chromosomes during meiosis resulting in recombination of genes.

B. Positional cloning can be employed to locate a disease-causing gene on a particular chromosome when the mutant allele (x) is inherited along with a particular set of genetic markers not associated with the normal nonmutant allele.

studied included characteristics such as length of the plants' stems, the color of the flowers, and the shape of the seeds.

Plants have sex, sort of. They pollinate. Most flowering plants are hermaphrodites that produce both male and female gametes, i.e., sperm and egg. Such plants can either self-pollinate, combining their own male and female gametes, or cross-pollinate, swapping genetic code with other plants of the same species. (Whether plants enjoy one method more than another is unknown.)

Mendel developed very careful procedures to ensure controlled fertilization so that he could experiment with both cross-fertilization and self-fertilization. He then would note the frequency of the appearance of specific traits. Mendel noticed that sometimes when he crossed two plants with the same version of a trait, say long stems, he ended up with 100 percent long-stemmed offspring. Other times, some of the plants would have short stems. Further, sometimes the offspring of self-fertilized plants had traits different from the parent plant. Mendel did these experiments many, many times and kept very careful counts. He found that sometimes when he planted seeds from self-fertilized plants, 75 percent of the offspring would have the parental version of the trait, while 25 percent would have the alternative.

For example, if you planted 100 seeds from a self-fertilized pea plant with a long stem, you might expect all the new plants to have long stems, like the parent's. This did happen with some long-stemmed plants when they were self-fertilized. Surprisingly, however, when other long-stemmed plants were self-fertilized, about 75 of the plants would have long stems and the other 25 short ones. That meant there were two different kinds of long-stemmed plants, even though from the outside they looked identical. If one of the long-stemmed plants that, when self-fertilized, resulted in the 75:25 split was crossed with a short-stemmed plant, about 50 percent of the offspring would have long stems and the other half short. Alternatively, if you crossed the type of long-stemmed plant that, when self-fertilized, resulted in 100 percent long-stemmed offspring with a short-stemmed plant, you would still end up with 100 percent long-stemmed plants.

From these observations, Mendel determined that long stems were somehow preferred when a long-stemmed plant was crossed with a short-stemmed one. Medium-length stems would generally not be observed in Mendel's crosses. Rather than an intermediate phenotype arising, the long stems were somehow "dominant" over short stems. The phenotype is basically the physical manifestation of the expressed DNA, or genotype, as modified by genetic, environmental, and epigenetic forces such as gene silencing by DNA methylation. Such means of phenotypic regulation cause variations that aren't accounted for in Mendelian theory. They are, for example, why identical twins and other kinds of clones are never perfectly identical.

When Mendel did his crosses, short-stemmed plants could only result if two copies of whatever made stems short were crossed. Nowadays we refer to the gene that controls stem size as having two different alleles, with the long allele "dominant" and the short allele "recessive." A plant with two copies of the same allele (both dominant or both recessive) is referred to as homozygous; if it has one of each type, it is called heterozygous. A homozygous-dominant plant will look the same as a heterozygous plant—that is, both will display the dominant trait, in this case, a long stem. Only when a plant is homozygous-recessive will it display the recessive version of the trait. So when Mendel self-fertilized a plant with long stems, if that plant was homozygous-dominant, 100 percent of the offspring would have long stems. However, if the parent was heterozygous, containing one copy of the allele for long stems and one copy of the allele for short stems, 75 percent of the progeny would have long stems and the other 25 percent would have short stems.

To visualize this, let's refer to the dominant allele for stem length that results in long stems with an uppercase **L** and the recessive one for short stems with a lowercase **l**. Following self-fertilization, a long-stemmed plant that is homozygous-dominant, **LL**, will only have offspring that are LL and they, of course, will all have long stems. However, for a plant that is heterozygous, **Ll**, there are four ways in which those alleles can be combined, which means that in 100 offspring plants, the distribution of alleles in the offspring will look like this, or very close to it:

25 LL	25 Ll
25 lL	25 ll

Only the 25 plants that inherited two recessive alleles (**ll**) would have short stems. Of course, Mendel had no way to know whether a particular long-stemmed plant was homozygous-dominant or heterozygous, but the results of a self-cross would tell him.

Another hugely important observation Mendel made was that two different traits, such as stem length and flower color, were not usually linked together. This meant he could separate out the genetic basis for different traits and study them independently. This led Mendel to what is referred to as *the law of independent assortment*. Basically, this law means that if your brother has the same curly hair as your mother, that doesn't mean he must also have her blue eyes. Or, inheriting your father's height doesn't necessarily mean you will also have his receding hairline.

Phenotypic traits—genetic characteristics as exhibited in individuals—generally are independent of one another. Of course there are genetically based variations in phenotype that are displayed with multiple traits, such as red-haired people often having freckles, but that is because there is a common genetic basis to the two traits. Whether a pea plant has a long or short stem has nothing to do with whether it has a white or purple flower. This was the case for the different traits Mendel studied and is generally true. The law of independent assortment holds—except when it doesn't.

When gametes—sperm and egg—are formed during the process of sexual cell division (meiosis), something very strange happens. You will recall that chromosomes are chains of DNA generally millions of nucleotides long. Humans have 46 chromosomes. Pea plants have 14. Chromosomes are paired, with one in each pair coming from the egg/mother, and the other from the sperm/father. Pea plants have 7 pairs and humans 23. As we know, along a chromosome, DNA is separated into genes, which are transcribed into mRNA, which is then translated into proteins. There are subtle variants of genes that exist in a population, referred to as alleles. Often, the two alleles of each gene that an individual possesses aren't exactly the same. These different alleles can result in differences in traits that are apparent in the phenotype of the new organism.

There is a mechanism that causes alterations in the specific complement of alleles carried by individual chromosomes. During a particular phase of meiosis, when gametes are being formed, the two pairs of each chromosome type line up side by side. They then start flopping over each other. Pieces of one chromosome stick to the adjacent paired chromosome in a process referred to as "crossing over," and this leads to the process of recombination, in which the two paired chromosomes exchange some of their alleles along with adjacent pieces of DNA. Aspects of this process were first described by Barbara McClintock, who, as you have read, won a Nobel for her research on transposons. Recombination changes which alleles of each gene are found on individual chromosomes. This increases genetic diversity, even when reproduction occurs by self-fertilization.

At a later stage of meiosis, individual gametes are formed. These have half the number of chromosomes normally found in the organism, one from each chromosome pair. So each sperm and egg are formed *after* the crossing over and recombination phase, and this generates a wide variety of possible combinations when fertilization ultimately occurs.

For example, imagine that the two chromosomes, call them A and B, each contain a different allele for three different genes. So you have one chromosome with the alleles 1A, 2A, and 3A and another with 1B, 2B, and 3B. Via recombination, exchange of genetic information might result in one chromosome with 1A, 2B, and 3A, and the other with 1B, 2A, and 3B. However, crossing over and recombination can happen differently every time, so many different combinations of alleles are possible. Thus, you can also get 1B, 2A, and 3A or 1A, 2A, and 3B. During fertilization, any of these different variants can fuse to form the offspring. A sperm with 1A, 2A, and 3A fertilizing an egg with 1B, 2B, and 3B would result in an exact clone of the parent plant, but crossing over and recombination make the chance of this nearly impossible. More likely, even in the case of self-fertilization, each seed will contain a different combination of genes, depending on the recombination that occurred during formation of each gamete.

Crossing over and recombination increase genetic diversity, as the different alleles of a gene are mixed up during gamete formation. While the law of independent assortment generally holds for individual traits, it is not 100 percent accurate, as it gives the impression that just the single allele undergoes crossing over. Recombination is not an exact process, so any DNA that is very close to a gene will be swapped over as well. Sometimes this means that certain alleles that are very close to each other on a single chromosome are almost always inherited together. This is referred to as linkage. However, given the total number of genes, about 20,000 in humans, the odds of any particular two being linked is very low. On the other hand, there is a lot of DNA in our chromosomes that does not encode a gene, and just having a small amount of adjacent "noncoding" DNA linked to an allele can be a powerful tool. Over the past few decades, geneticists have recognized this factor and made use of it in hunting for genes that cause disease.

One allele can encode for one variant of a phenotypic trait, and this includes genetic diseases. Such diseases are often recessive. Thus, if both parents are carriers for the disease, heterozygous for the disease gene variant but phenotypically normal, only 25 percent of offspring will be afflicted with the disease. This is of course limited to monogenic diseases, those caused by mutations in a single gene that is inherited as a conventional recessive trait. An example of this is cystic fibrosis, which leads to early death due to lung problems. Recessive genes can often lead to death prior to the age of reproduction, which is tragic for the individual, but limits the chance of the gene being passed on. It's important to understand, though, that many of the most prevalent diseases, such as heart disease, are caused by multiple genetic and environmental factors, and thus cannot be understood through simple Mendelian genetics.

This leads us to a question that is becoming a big factor in modern medicine. How do we identify the gene responsible for a disease?

There are a number of reasons why it might be important to know which mutated gene is responsible for a particular disease. It allows people in an affected family to be screened for status as potential carriers. It can help guide development and selection of specific therapeutic options that might be useful for one genetic basis of a disease, but not another.

Consider a patient who is diagnosed with a genetic disease. Geneticists figured out that if they could get DNA samples from the patient *and* a large number of people in the patient's family, they could determine the location and identity of the guilty gene. The family group would include healthy, unaffected people who were either homozygous for the dominant, normal allele or heterozygous carriers of a single copy of the disease-associated allele. The patient and any other relatives who had the disease would be homozygous-recessive. As described earlier, when crossing over happens, some flanking DNA around the allele in question is also recombined onto the adjacent paired chromosome. These innovative researchers determined that if they knew the exact sequences of many small pieces of DNA scattered throughout the genome, which are referred to as "markers," they could discover the location of the mutant gene—where it lies on which chromosome—and its identity. With a sufficient number of markers, hundreds or even thousands, all you have to do is figure out which genetic markers each member of the family has and then see if any particular markers are linked to the disease, which is to say, always inherited along with it. As the markers correspond to known locations on particular chromosomes, this gives you a very good idea where the disease-causing gene can be found. As long as you have enough family members—affected, carriers, and normal—and a large number of markers that vary in the population, some of which will be linked with the mutant variant that causes the disease, you can identify the location of the gene.

Back in the days before the human genome had been entirely sequenced, this kind of search could be arduous. Just knowing that the disease-causing gene was located relatively near a particular genetic marker on one part of a specific chromosome didn't tell you the exact identity of the gene. The human genome was sequenced a bit at a time, so a researcher might get lucky and find that a known gene that had a function relevant to the disease being studied had already been identified in the area near the marker of interest. If not, the researcher had no choice but to start sequencing DNA, beginning at the site of the marker and looking in both directions along that chromosome. Although DNA sequencing is becoming easier and less expensive, it can still be costly and time-consuming if you need to cover thousands of nucleotides. However, now that the entire human genome has been sequenced, we essentially know the location and identity of every human gene. Thus, we can now use the genome as a map and the genetic markers as pins. We identify the pin that is linked to the disease we are studying and then look on the map to see which genes are nearby. Say, for example, that we are looking for a gene that causes an inherited neurological disease. If we find, near the marker, one, and only one, gene that is known to be expressed in the brain, we are well on our way. Of course, ultimately, the potential disease-causing variant of the gene will need to be sequenced and compared to the normal allele in order to verify that a mutation exists. The first human genome sequence cost an estimated $3 billion. Currently, the price per genome is in the thousands of dollars. So, although the basis of many monogenic diseases, such as cystic fibrosis, have been determined through linkage with genetic markers, it may be that quite soon all genetic questions will be answered through quick, cheap, and easy analyses based on full-genome sequencing.

CHAPTER 7

The Revolutionary Reaction, or, How to Make DNA in Your Kitchen

◉

W E LIVE IN AN AGE OF GENETIC WONDERS. SCIENTISTS have sequenced the entire human genome. We can identify mutant genes, test for viral diseases, perform genetic engineering, and help solve crimes and understand more about people who lived hundreds or even thousands of years ago. At this point, you might be asking how scientists have done this and how they make DNA when they need it.

The answer to these questions is the polymerase chain reaction (PCR).

PCR is quite simply the most revolutionary experimental technique ever developed in the history of molecular cell biology. And while you probably can't really do it in your kitchen, every single graduate student in the field has done it in a lab, and every lab has a PCR machine. Now even robots are doing it.

In order to understand how PCR works, we must first learn a little bit more about how DNA replication, the copying of specific segments of DNA, occurs.

The polymerase chain reaction (PCR), or how to duplicate
a specific piece of DNA. (1) Heating DNA to 95°C leads
to separation of the two strands of the double helix. (2)
Cooling to 60°C allows binding of the primer. (3) Warming
to 72°C leads to DNA polymerization.

In all organisms, from bacteria to us, DNA replication follows
the same basic formula. You will recall that DNA is made of two
strands, a double helix, linked together by bonds formed between the
paired nucleotides across the rungs in the ladder. This is referred to
as Watson-Crick base pairing. The first step in DNA replication is
to separate the two strands. In our cells, this is triggered by helicase,
so named because it is an enzyme that breaks apart the double helix,
leaving two single-strand DNA polymers.

In science, suffixes often provide insight into function; "-ase" des-
ignates an enzyme. An enzyme is a protein that does some kind of
work; it accomplishes some sort of chemical change. Enzymes are
not simply structural components, part of the building's framework.
Rather, they are molecular saws and nail guns, actively altering the
molecular and cellular structure or function that surrounds them.

Once helicase does its job, the DNA polymerase—the enzyme
that makes more DNA polymers—can start copying each of the
individual template strands, generating two new double helixes.

As we learned from Meselson and Stahl, DNA replication is semi-conservative: once helicase separates the parent double helix, each strand will have a newly synthesized strand added to it. This will result in two new double helixes, each with one original parental strand and one new daughter strand.

In order to do its job of replicating a single-strand DNA polymer, DNA polymerase needs help from other factors—including a source of free nucleotides and a "primer." A primer is a short stretch of polymerized nucleotides that is an exact complementary match to the beginning of the sequence you want to replicate. Primers only 20 nucleotides long can be sufficient to provide a specific starting point for the replication reaction. Only after the primer binds to the template strand (the one being copied) via Watson-Crick base pairing can DNA polymerase begin to grab free nucleotides and put them in line in the correct order (complementary to the rest of the template), forming a phosphodiester bond between each subsequent nucleotide like links in a chain. At the same time, the bonds form across the paired bases, making up each rung of the DNA double-helix ladder.

For a very long time, scientists have used bacteria as little DNA factories. If you can introduce a piece of DNA into bacteria, preferably within a circular piece of DNA referred to as a plasmid, you can trick the bacteria into synthesizing new copies of your DNA of interest. Bacterial DNA polymerase likes to begin its synthesis reaction at what is known as an "origin of replication," a specific 75-nucleotide sequence that binds proteins involved in the replication reaction, such as helicase, and allows the process to begin at a particular location. Thus, you need to make sure the plasmid containing the piece of DNA you want to duplicate contains an origin of replication.

It is actually quite easy to get external DNA into bacteria, in particular a lab-adapted strain of *E. coli*. The process is referred to as transformation, and all it requires is warming the bacteria up a few degrees in the presence of the circular DNA plasmid of interest. In addition to an origin of replication, most scientists will include a gene to encode a protein that will make the transformed bacteria resistant to a specific antibiotic, such as ampicillin. Generally, this is accomplished with a gene that encodes a protein that pumps specific antibiotics out of bacteria. Then you grow the bacteria in the presence of the antibiotic, so that any bacteria that were not successfully transformed will be killed.

The number of cells in a culture of *E. coli* can double in as little as twenty minutes, so if you grow your transformed bacteria overnight, you can end up with a lot of bacteria that have replicated your plasmid. With a doubling time that rapid, it actually takes only ten hours to go from a single bacterium to a billion. In many cases, genetically engineered mutant origins of replication are introduced into the plasmid to trick the bacteria into creating tens or even hundreds of copies of the desired DNA per cell, speeding DNA production even further. Even so, this process is a lot of trouble if you just want to make copies of a specific stretch of DNA, especially as you still need to purify the plasmid DNA of interest away from the rest of the bacteria. That is where PCR comes in.

About thirty years ago, a biochemist named Kary Mullis was thinking about potential ways to amplify specific DNA sequences without using bacteria or any other cells. Mullis was driving around with his girlfriend when the idea came to him. This was one of those rare examples of a true "eureka" moment in science. He reasoned that if you had two primers that would bind on the opposing strands of a specific sequence of DNA—one at the start of the sequence you want to duplicate on one strand and the other at the end of the sequence that binds to the opposing strand—and would allow two synthesis reactions to occur moving from one primer to the other, and if you could do this over and over again, you would end up with many copies of the sequence between the two primers. The problem was how to get this to actually work in practice.

Mullis had the recipe: you needed your target template sequence, free nucleotides, primers complementary to separate stretches along the parent double helix (one on each strand), and DNA polymerase. But how would you break the Watson-Crick base pairing holding the double helix together? If you added helicase enzyme, it would prevent the reaction from working, because it would be constantly separating out adjacent strands. Mullis realized that heat could do the work. If you heat up DNA to just below the boiling point of water, around 95°C, Watson-Crick base pairing falls apart and the double helix separates into the two separate strands. And, happily, this is reversible. When you cool the DNA mixture down again, the complementary strands zip right back up. But if you add a large excess of primer to the mixture, then, as it cools, the individual parent strands are much more likely to bind to a primer molecule than to find another full-length complementary strand to pair with. This meant that, in the presence of DNA polymerase, you could synthesize the desired new chunks of DNA along each template between the primer molecules.

However, one problem remained. Although DNA can withstand heating to high temperatures, the DNA polymerase that Mullis initially was working with could not. Proteins generally "denature" irreversibly at high temperatures, basically turning into scrambled eggs. This meant that Mullis had to add the DNA polymerase only after the reaction had been cooled down, to allow primer binding to the template strand. This was tricky and inefficient, as each time you heated the reaction, you would kill any polymerase present. However, eventually Mullis and his colleagues had a truly ingenious idea: they simply looked for and found a DNA polymerase that can withstand high temperatures.

This is not as crazy as it sounds. There are organisms that can live in extreme environments. In 1976, scientists who study these extremophiles isolated a DNA polymerase from the bacteria *Thermus aquaticus*, which lives in hot springs and geysers and can withstand near-boiling temperatures. The DNA polymerase from *Thermus aquaticus*, Taq polymerase for short, was the key addition to Mullis's dream reaction. In the presence of the template, primers, nucleotides, and Taq polymerase, the reaction could be heated to 95°C to separate the strands of the double helix, cooled to approximately 60°C to allow Watson-Crick base pairs to form between the individual template strands and the primers, and then finally heated to 72°C, at which temperature the Taq polymerase will synthesize new DNA polymers off of the primers with great efficiency.

Repeating this thermal cycling, 95°C to 60°C to 72°C, again and again creates a chain reaction that greatly amplifies the starting template specifically at the region between the two primers. In the laboratory, thirty cycles will usually be run, each total cycle lasting only a few minutes. That means in a few short hours, a single piece of DNA can be copied approximately half a billion times. Mullis started out using a stopwatch and water baths set to different temperatures. Today we have sophisticated PCR machines—computer-controlled thermal cyclers that hold thin-walled plastic reaction tubes and allow for very precise and rapid heat transfer.

PCR is routinely used in nearly every facet of biomedical science and is an absolutely essential part of every laboratory. PCR is integral to various methods for DNA sequencing and essential for all types of genetic engineering. PCR is used to introduce specific mutations into DNA, for example by synthesizing primers that mismatch the template at a single position. It is used to amplify trace DNA found at locations ranging from crime scenes to archeological digs. PCR is used to identify which genetic markers an individual carries in linkage studies to identify disease-causing genes. It can also be used as an extremely sensitive test for infection, for example to see if viral DNA, such as from HIV, is present in a blood sample. Thus, PCR truly reigns as the most notable technical advance in molecular-cell biology, and for his discovery, Mullis was awarded the Nobel Prize in Chemistry in 1993.

CHAPTER 8

Piecing Together the Puzzle: How We Sequence DNA

A S STATED IN THE PREVIOUS CHAPTER, THE POLYMERASE
chain reaction (PCR) lies at the core of many methods
for DNA sequencing. Although numerous specific techniques have been developed for deriving the order of nucleotides in a
stretch of DNA, a few have emerged as the most efficient and cost-
effective. While a few short decades ago most sequencing was performed manually and was limited to short stretches of only a few
hundred nucleotides, current technology allows the rapid and inex-
pensive sequencing of whole genomes with next-generation machines
that can cost nearly three-quarters of a million dollars, not including
the "consumables" costs associated with running the thing. In many
techniques, the larger sequences—such as whole chromosomes, which
run millions of nucleotides long—can be constructed by treating a
single, long piece of DNA as if it were simply an assemblage of many
short pieces. However, many of these different more-complex methods
still fundamentally rely on the method developed by Frederick Sanger
that won him the Nobel Prize in Chemistry in 1980. This was actually
Sanger's second Nobel; the first was awarded in 1958 for his work
on protein structure. Sanger remains the only person ever to win two
Nobel Prizes in the same category.

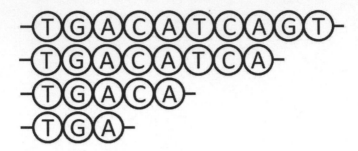

DNA sequencing by the Sanger chain-termination method.

The method for DNA sequencing that Sanger first developed was called the "plus-and-minus" protocol. This involved a very complex and time-consuming series of steps that included replicating the DNA template strand, adding and removing different combinations of nucleotides, followed by biochemical purifications and analytical approaches. The general idea was that in the absence of one of the four nucleotides that made up DNA, the replication reaction would stop. If you could tell the length of that arrested replication product, not easy in the mid-1970s, you could figure out the position where the nucleotide normally went. This was very painstaking and difficult and only applicable to very short sequences, no more than about fifty nucleotides long. However, this led Sanger to a variation on the same general theme that would prove much more rapid and efficient and that is still in use today.

Imagine that you are making beaded necklaces with four different types of beads labeled A, T, C, or G for the four nucleotides. All the necklaces have ten beads in the same order, TGACATCAGT, and you have thousands of them. Now imagine that someone who is blindfolded wants to know the order of the beads in your necklaces, and you can't talk. The blindfolded person hands you a scissor and tells you to cut the string after the As, but only some of the As and on only a few necklaces. If you cut after one or another of the As in about 1 percent of the necklaces, you would end up with a few necklaces cut after the first A, a few after the second, and a few after the third, plus a large quantity of uncut necklaces. A few would read TGA, some TGACA, and a few TGACATCA. So the blindfolded person could count the length of these and determine that there were three As in the sequence at positions 3, 5, and 8.

If you repeated this exercise cutting necklaces after Ts, Cs, and Gs, eventually the blindfolded person could figure out the exact sequence of all ten beads in the chain. This is exactly how Sanger sequencing works, except, obviously, we don't use scissors. Instead, a small proportion of the one nucleotide at which we want to "cut" is chemically modified so that, once it is added to the DNA polymer, that chain can't be extended beyond it. This is why the other name for Sanger sequencing is the "chain termination" method. If four separate such reactions are performed, one for each nucleotide, you can eventually determine the entire sequence—if you can precisely measure the length of all the DNA molecules.

The way these special chain-terminator nucleotides work is simple. As links in the DNA polymer chain, each nucleotide has to participate in two bonds, one with the nucleotide in front of it and one with the nucleotide after it. Imagine that DNA is normally like a series of chain links that consist of a metal rod with a ring at each end, like this . . .

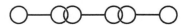

. . . but that the special chain-termination nucleotides are missing the ring at one end, like this:

When one of these is added, the chain ends, because the next link has nothing to which to attach itself. Again, the key would be to have only a small amount of terminator links in the presence of many normal, complete links, and a large number of forming chains, so that you end up with chains of all different possible lengths.

Which brings us to this question: How do we figure out the exact length of a piece of DNA?

The answer is basically electrified Jell-O.

Agarose is a gel that is made from seaweed extract. As a gel, it has a lot of water in it, so solutions can be added, but there are also holes through which molecules can move—so it is sort of like a molecular sponge. DNA is negatively charged due to its phosphate groups. So we start with a rectangular slab of agarose. We add DNA to one end, and apply an electric field with the negative charge at the same end as the DNA and the positive charge at the other end, which causes the DNA to move toward the positively charged end, as it is repelled by the negative charge and attracted to the positive. This is called electrophoresis. The trick is that small pieces of DNA will move more easily and quickly through the gel than larger ones; therefore, with a mixture of DNA polymers of different specific lengths, we end up with a series of bands where the DNA of different specific sizes has migrated together. If we stain the gel with a fluorescent DNA-binding dye, we can see the bands containing each group of DNA molecules of a different particular size.

When this method was first being used, you had to prepare four different samples, each with a small amount of one specific chain-termination nucleotide (A, T, C, or G) and run reactions for each on separate gels. You ended up with four different patterns of bands that had to be visually inspected to figure out the template sequence from the different lengths of the DNA molecules made with all the separate combinations of chain terminations. Because this was all done manually, and bands might differ in length by as little as one nucleotide (where a sequence contained a doublet of AA, TT, CC, or GG), the gels used for sequencing were huge, the size of a cookie sheet so the tiny differences in length could be seen, and it took great skill to prepare the tests and analyze the results.

The modern, mature form of Sanger sequencing has become automated and much simpler. The four different chain-termination nucleotides are used in very small proportions to the bulk normal nucleotides. However, each has been labeled with a different color. That way they can all be mixed together in one reaction. PCR is used as a way to rapidly make many copies of the template, some incorporating a chain-termination nucleotide of each kind—A, T, C, or G. These different PCR products are then all analyzed together with a machine that can very precisely and rapidly tell the order of colors in the group of mixed DNA molecules following electrophoresis.

While this represents a big increase in speed and efficiency, Sanger sequencing only works for DNA segments of up to about one thousand bases, since that is the longest PCR product we easily produce. So how do we sequence whole chromosomes that are millions of nucleotides long?

Today we have a series of complicated next-generation sequencing procedures and instruments that employ innovative techniques for obtaining very long DNA sequences. But the human genome was first sequenced the old-fashioned way—by a huge number of conventional Sanger sequencing reactions. The Human Genome Project involved thousands of researchers around the world collaborating to produce a complete sequence of all human genes. This was accomplished by breaking the 46 chains of millions of nucleotides each that make up the human genome into many smaller pieces that could individually be subjected to Sanger sequencing.

To break up these chains, we use "restriction enzymes." These are essentially components of the bacterial immune system. Bacteria protect themselves from phage, the simple viruses that infect them, by expressing a series of DNA-cutting enzymes that have evolved to specifically target phage sequences. The bacteria use these to slice up the phage DNA before infection can occur. These "restriction sites" exist throughout our genomes as well, not surprising considering the human genome is billions of nucleotides long and contains integrated sequences from a variety of pathogens going back in time to our evolutionary forebears. If you want to cut up DNA, all you have to do is mix it with the right combination of restriction enzymes and, snip, snip, snip, you are left with a bunch of small fragments that can be sequenced, like cutting up fresh noodles into individual slurpable pieces.

Once you have sequenced all these small pieces, you do then have the problem of putting them back together to get the whole picture again. The solution to this is to make and sequence many different versions of the cut-up genome. When assembled together, these small sequences overlap each other. Imagine that you had a DNA molecule two thousand nucleotides long and you broke it up into a series of different-sized pieces that were not mutually exclusive, that is to say you had different subpopulations within the mixture. If some of the pieces contained nucleotides 1–800 and others nucleotides 500–1200, and some 1000–1600, and then a group with 1500–2000, then, because of the overlap, you could rearrange them all in the correct order once each fragment had been sequenced. The process of breaking up large stretches of DNA like this is called "shotgun sequencing," because you blow the DNA template up into small fragments. It is very powerful, but extremely time-consuming and requires computer programs in order to stitch everything back together. Also, scientists in the field of bioinformatics are involved in determining how much starting material you need, how many fragments you should create, and how it all gets organized at the end. This is also very expensive. While the first human genome cost billions of dollars, with the latest technology this price tag has dropped into the thousands. Thus, we may be entering an era when each person's individual genome can be sequenced for about the price of an MRI scan. This is rapidly paving the way to "personalized medicine," a catchy phrase that is increasingly uttered by those involved in developing new diagnostic and therapeutic approaches and will likely lead to very different types of medical treatment strategies in the near future.

CHAPTER 9

The Genome
and Personalized Medicine:
Progress, Promise, and
Potential Problems

W E NOW LIVE IN A WORLD IN WHICH THE ENTIRE HUMAN
genetic code—the genome—has been sequenced gene
by gene, and in which every individual's own personal
genome can easily be sequenced as well. This presents new risks and
benefits, and it's important to understand what this means to us.

Actually, a database of full-genome sequences is overkill for most
applications. Our 23 pairs of chromosomes contain more than three
billion base pairs constituting 20,000 to 25,000 genes. Yet, when
we're looking for a particular set of genes for, say, medical or forensic
reasons, we don't need to go to the time and expense of sequencing
the person's entire genome.

Using restriction enzymes to cut up a sample of someone's
DNA, as described in the previous chapter, will give a unique pattern when the resultant fragments are visualized on an agarose gel.
This is because no two genomes are exactly alike, except with identical twins. Polymorphisms are the naturally occurring variations
in the genome from one person to the next. The word *mutation*

DNA sequencing of a tumor can lead to targeted therapy that can selectively inhibit the mutant protein responsible for the tumor. However, other mutations present in some cells can result in further tumor growth.

suggests either a particular benefit, as in Darwinian natural selection, or disadvantage, such as in causing disease. But polymorphisms are just differences and do not denote any specific outcome of those changes. Because of polymorphisms and other allelic variations within and between populations, no two genomes are exactly the same. Thus, restriction fragment patterns are as unique as fingerprints and have been used for decades in paternity cases and criminal investigations. With full genome sequences, we can certainly learn a great deal about complex diseases that emerge from multiple genetic and environmental factors, yet we are currently very far away from identifying at birth who will develop heart disease or Alzheimer's. So, while we can sequence full genomes, there are currently few uses for that information.

The ability to rapidly and accurately sequence specific individual genes by PCR reactions following the modern Sanger technique is extremely powerful, though. We can screen for many types of genetic diseases caused by particular mutations in specific genes. While it is true that having a reference genome sequence, which tells us the identity and chromosomal location of every human gene, is a huge advancement, the ability to sequence each individual's unique genome with all its many variations, polymorphisms, and mutations has not yet resulted in many clinical benefits. Even though we are nearing the point where obtaining an individual's complete genome sequence could be a medical and economic reality, there is little benefit in doing so, when compared with more targeted approaches.

The general concept behind "personalized medicine" is that the more the doctor knows about a patient's genetic background, the better the physician can tailor treatment and prevent or manage future problems. One medical perspective in favor of whole-genome sequencing is the concept of the biomarker. This suggests that any specific, quantifiable characteristic associated with a particular disease can be used to identify that disease before any symptoms are evident. Biomarkers have been sought for many different diseases from a wide variety of sources and specific types of molecules. This has been tried, generally unsuccessfully, with many different biological materials, including blood, urine, hair, and pretty much anything else doctors and scientists can get their hands on. As the genome is all the DNA that makes up our genes, the "transcriptome" is all the mRNA that encodes proteins, which when analyzed together are referred to as the "proteome." Similarly, the "metabolome" is the sum of all the different types of metabolic factors such as sugars and vitamins in our bodies. Genomics, transcriptomics, proteomics, and metabolomics are all being employed to develop diagnostics and guided therapies from the point of view of biomarkers. While there certainly will be some successes in these endeavors, none will be a panacea that allows accurate predictive diagnosis of all diseases that individuals might develop over their lifetimes, or the ability to develop personalized therapeutic regimes for every individual.

There are, of course, victories for the proponents of personalized medicine. There are people who cannot metabolize certain drugs, which makes those drugs toxic to them rather than therapeutic. Knowing who will respond to a particular therapy before giving the drug can be a huge benefit to doctors, and DNA sequencing can provide insight into the design of an optimal therapeutic regimen. This is particularly relevant in the case of cancer treatment.

Established chemotherapy agents kill rapidly dividing cells indiscriminately, while causing significant side effects. Doctors are trying very hard, and succeeding in some cases, to replace these with therapies specifically targeted to fight cancer cells that possess particular mutations.

There are genes called oncogenes, which encode proteins that normally drive cell proliferation, but when mutated to be overly active can cause cancer. Many oncogenes are involved in the ability to sense and convey signals that the body sends to tell cells when tissue growth is needed. These growth-factor signaling pathways are very common places for cancer-causing mutations in oncogenes to arise. The result is as if the cell is running a race on amphetamines. With the development of cheap and easy DNA sequencing, it is possible to identify the mutations in an individual tumor and employ drugs that will act specifically to inhibit the oncogene that is causing the problem. There are many enzymes that will modify proteins in the signaling pathways that affect cell proliferation. It's sort of like a game of telephone at the molecular level. The growth factor binds its receptor on the cell surface, which activates a cascade of enzymes that modify each other in series, ultimately resulting in cell proliferation. However, when one of these enzymes is mutated, the oncogene acts as if a person in the middle of the line started the game of telephone without anyone else before him or her saying anything. This leads to cell replication at inappropriate times—in a word, cancer.

Biochemists have had great success developing selective inhibitors to the various enzymes in these growth-factor signaling pathways, including agents that selectively inhibit only the oncogenic variants of those enzymes found mutated in cancer. One of these signaling enzymes is called RAF, which stands for rapidly accelerated fibrosarcoma. Sarcomas are cancers in connective tissue, and they can be caused by viral infection. In fact, RAF was originally discovered in a virus that causes sarcoma in mice. Why would causing cancer benefit a virus? The only goal of a virus is to cause a cell to make more virus. If a virus can stimulate infected cells to proliferate, the result will be more potential for virus production. More cells, more virus. The viral version of RAF is more active than the normal cellular version in humans and mice. It stimulates cell proliferation as an oncogene. Similarly, the normal version of RAF becomes more active when mutated, and this mutant version of RAF has been the target of specifically developed inhibitors.

One specific hyperactive mutant of RAF, in which the 600th amino acid in the protein has been changed, is found in 7 percent of all human malignant cancers and 60 percent of melanoma skin cancers. Vemurafenib is a drug developed to inhibit this mutant RAF without affecting normal RAF in noncancerous cells. Thus, before administering this drug, doctors must perform DNA sequencing on the patient's tumor to determine if it is in fact carrying this specific RAF mutation. If not, there is no point in administering Vemurafenib.

Although immediate benefits were seen following treatment with Vemurafenib, in many cases the cancers eventually returned. Further studies revealed what allowed secondary tumors to form despite the treatment. Basically, the cells in the tumors that formed in the relapse were driven by mutations in proteins other than RAF. This seems a clear case of Darwinian natural selection on a cellular scale. Cancer cells are genetically unstable, which results in rates of mutation far beyond normal. So if you have a mixed population of cancer cells, some with the RAF mutation and some without, and you eliminate the RAF mutant cancer cells, the other cells that remain will eventually thrive and grow into tumors.

While Vermurafenib is an example of the frustrations that sometimes plague the use of personalized medicine for cancer treatment, not all attempts have been as difficult.

Some breast cancer cells can express the receptor for the hormone estrogen; others don't. These are referred to as estrogen-receptor positive or negative tumors. About 60 percent of breast tumors are estrogen-receptor positive. The growth of estrogen-receptor positive tumors is promoted by estrogen in the patient, so women with breast cancers have their tumors tested to see if they are estrogen-receptor positive or negative. If the tumor is estrogen-receptor positive, then the patient will be given so-called hormone therapy. Treatment with the drug Tamoxifen blocks the ability of the estrogen receptor to respond to estrogen, thus halting the proliferation of estrogen-receptor positive cancer cells. Clearly, we are only scratching the surface of the potential for personalized medicine. However, in a world where extending even basic conventional treatments to all has not been economically possible, it may not really be ethical to invest heavily in diagnostic and treatment options that, at least for the foreseeable future, might only be useful for a small number of rare diseases.

CHAPTER 10

The Science, Technology, and Ethics of Manipulating the Genome

T HE ABILITY TO SEQUENCE INDIVIDUAL GENES, OR EVEN the whole genome, can provide tremendous analytical power, but what about the next step—actually changing the genetic code in a living organism? The ability to change or deactivate a gene, for example one that causes a genetic disease, could lead to therapeutic benefits. That being said, making changes to the human genome, especially ones that could be passed on to future generations, raises significant ethical questions. Moreover, actually accomplishing this is far easier said than done.

Although bacteria can be easily transformed with plasmid DNA, it is much more difficult with a multicellular mammal like us, or even a model organism like a fruit fly or a mouse. During the process of meiosis, when gametes (sperm and egg) are being formed, the two copies of each chromosome line up and exchange genetic material to increase diversity in potential offspring by the process of crossing over. This is the same general principle by which the genome might be altered in a laboratory, once fertilization has occurred. One example of this that has been incredibly powerful

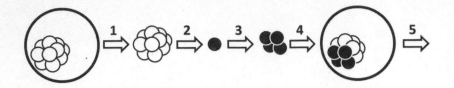

How to make a knockout mouse in five easy steps: (1)
Embryonic stem cells are removed from the inner cell
mass of a blastocyst (early-stage embryo). (2) The knockout
construct, which contains a nonfunctioning piece of DNA
partially identical to the gene of interest, is introduced.
(3) Cells that have successfully integrated the knockout
construct survive antibiotic selection. (4) These cells are
placed into another blastocyst. (5) This chimeric blastocyst
is implanted into a pseudo-pregnant mouse, and the pups
are bred until a complete knockout mouse is produced.

in the understanding of human biology and disease processes is
the technique of target gene disruption, or "knocking out" a gene.
So-called "knockout" mice have been essential in understanding
the function of particular genes, by showing us what happens when
a particular gene has been completely removed. Moreover, similar
techniques can also be used to create genetically engineered "trans-
genic" mice, which have an altered or mutated form of a particular
gene in place of the normal version.

How do we create a knockout mouse? Very soon after fertilization, embryonic stem (ES) cells are removed from the inner cell mass
of a blastocyst, which is the small sphere of newly formed cells that
arises soon after fertilization—an early-stage embryo. These cells are
then triggered to take up a prepared DNA construct that will swap
places with the mouse's homologous copy. There are a few key criteria required to produce a suitable knockout construct. It must have
enough of the gene of interest to be recognized as homologous by
the cell, but it must have a significant portion removed so that a
functional protein will not be generated. More specifically, both ends
of the gene need to be somewhat intact or the recombination won't

occur, so usually the middle of the gene is deleted. Sometimes a small amount of the beginning of the gene will end up being expressed as a truncated protein that can display some unpredictable functions. However, with proper design, there will be no gene product generated at all.

Also, the DNA construct being introduced into the ES cells needs to contain some kind of selectable marker. This is usually an antibiotic-resistance gene that will protect the cells that now have the knockout construct while the unwanted cells are killed off with an antibiotic drug. These knockout cells are then injected into a new blastocyst, which is implanted in a female mouse that has been treated with hormones to induce a pregnancy response so that the blastocyst will grow into a mouse pup.

However, that is not the end of the story. That first mouse pup is not a full knockout. You probably noted that the blastocyst will contain normal cells as well as knockout cells. Also, the odds that both copies of the gene were knocked out in the ES cells that survived selection are infinitesimally small. So when the pups are born, they will have some normal cells and some cells with one normal copy of the gene, and the other with the knockout, referred to as heterozygous null. Therefore, the next step is to start breeding the mice more or less the way Mendel bred pea plants. We begin by breeding these mice with normal mice. When mice with heterozygous-null sex cells are bred with normal mice, one in four of the subsequent pups will have all their cells heterozygous null for the gene of interest. By breeding these heterozygous-null mice with one another, one in four of those pups will be complete homozygous-null knockouts in all cells. Of course, if you've knocked out a gene so important that the embryo dies, you won't get a mouse to study, but if the heterozygous-null mice are viable, you can often learn a great deal about gene function from them. Thousands of knockout mice have been successfully bred, and they have provided insights completely impossible to obtain any other way.

On the other hand, many biologists want to study cells, not whole organisms, so there needed to be a way to make knockout cells.

Until recently, the common way was to start the process of creating a knockout mouse and then culture some of the knockout cells. Usually, mouse embryonic fibroblasts (MEFs) are employed, as they are easy to culture and can be obtained very early in development, which lets us study them even if a knockout phenotype is lethal. However, this is quite a lot of trouble to go through just to get some cells. Luckily, a revolutionary new technology has recently been developed that allows targeted gene disruption in cells and even model organisms with much less trouble than it takes to produce generations of knockout mice.

This new technique is called CRISPR/Cas9, but we will simply refer to it as CRISPR, and is based upon a bacterial system for cutting up specific DNA sequences, sort of like a gene-specific restriction enzyme (see chapter 8). The CRISPR system can be engineered to specifically target a gene of interest much in the way that homologous recombination does, only much more simply and more efficiently. The CRISPR-based technique, which is referred to as genome editing, has only been around for a few years, but it already seems poised to create the next methodological revolution in biology, allowing for cheap and simple manipulations of genomes in single cells and organisms. Companies are starting up that will custom design ready-to-use CRISPR reagents, which can be used in a lab to rapidly edit genomes at will. Already, researchers in China have reported the successful genomic modification in human embryos, and human-genome-edited cells were credited with saving the life of a girl with leukemia.

Recently, a group from Harvard University led by George Church reported using CRISPR technology to alter 62 different genes in pig cells. Why pigs? Because of the potential for transplantation. There are many more people on transplant waiting lists than there are organs available from donors. Pigs and humans are very similar physiologically, and scientists have long suggested that a liver or kidney from a pig could save the life of a human experiencing organ failure. We already have success in transplanting some tissue, such as heart valves, from pigs to humans. However, this is only done following chemical treatments that prevent adverse reactions. Rejection by the immune system isn't the only problem. The pig genome also contains viral sequences that can be activated to produce endogenous viruses. These are like super-transposons that can move from cell to cell. Porcine endogenous retroviruses, or PERVs, presented a significant hurdle in the potential use of living pig tissue in transplantation. Church's group used CRISPR to inactivate 62 PERV genes from pig cells, which effectively eliminated the production and transmission of PERVs to human cells. Thus, in the future, organs like kidneys and livers could be harvested from genetically modified pigs for transplantation into humans.

These are only first steps, and it isn't yet clear where this technology will lead us. One possibility, genome editing of a developing human embryo to achieve favored phenotypes such as those with higher IQs or athletic ability, is seen by many as ethically dubious at best. On the other hand, CRISPR technology might eventually be used for widespread human gene therapy. The ability to permanently modify the genome of someone suffering from a genetic disease has seemed just around the corner for decades. Although there have been a few successes, gene therapy has mostly failed to live up to its promise. Fundamentally, gene therapy simply means the use of genetic material, DNA or RNA, to modify an individual to treat a disease. Generally speaking, correcting a mutation to cure a genetic disease has to be a permanent fix, so the question becomes: How do we get external genetic material into a person's cells and make it stay there? The best answer we've found so far is viruses.

So far, gene therapy has primarily focused on using viruses to insert a correct copy of a gene into a person suffering from a disease caused by a mutation. However, there are numerous hurdles to successful gene therapy, such as getting the gene of interest into a suitable viral particle, getting that particle to safely infect the individual, and getting the correct copy of the gene properly expressed. Some viruses can only hold small amounts of extra DNA or RNA. Others will only infect certain types of cells. And sometimes the immune system will attack the virus. In 1999, a young man named Jesse Gelsinger died in an early gene therapy trial at the University of Pennsylvania that was testing fundamental aspects of a potential treatment for the rare metabolic disorder that Jesse suffered from. His death was due to an immune reaction to the virus used in the trial. This sparked a major review of gene therapy trial design and implementation, along with a lawsuit and lingering safety concerns. There have been successes in gene therapy, such as with blood diseases where the affected cells could be removed from the body, treated, and then returned, or in cases in which the engineered gene therapy virus could be infused directly into the bloodstream. However, there has been a huge amount of investment in gene therapy with very little clinical success to show for it. Therefore, as with personalized medicine, we need to consider the balance of cost and benefit, particularly in a time when access to even basic medical care is limited to those with the means to afford it.

CHAPTER 11

Science Fiction and Social Fiction: What Is and Is Not in Our Genes

T HE MOVIE *GATTACA* WAS RELEASED IN 1997, AT THE SAME time that the race to sequence the entire human genome was reaching a fevered pace. In the film's "not too distant" future world, in which DNA could be sequenced extremely rapidly, two general outcomes were revealed. First, it was trivially easy to positively identify a person, whether entering one's place of work or in the context of a criminal investigation. Second, ideal offspring could be selected through genetic consultation and manipulation of the genome. Neither of these things is possible today, but they are not too far off. However, even more significant to the narrative of the film were the psychological and societal consequences of living in a world in which one's personal genetic code could be public knowledge.

As humans, we all have 46 chromosomes, 23 from our mother and 23 from our father. Except for the X and Y chromosomes, which determine sex, these exist in homologous pairs, i.e., pairs that correspond gene for gene and are thus nearly identical in sequence. We have two different copies, or alleles, for each gene, one on each chromosome in a pair. The few differences between our two alleles can

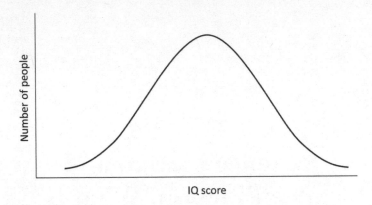

The authors of *The Bell Curve* discussed the distribution of IQs within the population and offered controversial opinions about social policy from what they saw as a genetically determined human trait.

be subtle, in which case they are referred to as polymorphisms, or severe, in the case of mutations. Let's call the two alleles of a specific gene A and B. A child can have any of four different combinations of maternal and paternal copies:

Mom-A/Dad-A
Mom-B/Dad-A
Mom-A/Dad-B
Mom-B/Dad-B

Thus, part of the conceit of *Gattaca* was that geneticists could aid prospective parents in selecting the "best" combinations of maternal and paternal alleles, those associated with, say, lowest chance of disease or highest intelligence. However, not all parents chose this route, and some children were still conceived the old-fashioned way. This led to two different classes in society, the genetically superior "Valid" people and the "Invalid" or "de-gene-erates." Most professional careers in the world of *Gattaca*, such as in law enforcement or space travel, were completely off-limits to those not deemed Valid.

However, through perseverance, determination, and a great deal of subterfuge, Ethan Hawke's character was able to attain a stellar trajectory, literally, and successfully impersonate a prospective astronaut. An equally compelling aspect of this film is the onus that several of the bona fide Valids feel based upon their supposed superiority. This leads to significant sibling rivalry with the brother of Ethan Hawke's character, who was genetically selected, as well as (spoiler alert!) the suicide of Jude Law's character, a genetically superior individual paralyzed in an accident. The concept that most but not all of the genetically inferior will be bent to the will of societal forces, while most of the elite will rise to lofty heights on the wings of expectation, are self-fulfilling prophecies that have clear parallels in our real, present world.

There is an interesting episode of *The Twilight Zone* from the early 1960s called "Eye of the Beholder" that highlights the arbitrary nature of societal expectations and conformity. In this world, a woman with what we would see as a beautiful face is identified as horribly disfigured in an alternate universe where most people have twisted lips and pig-like snouts. "Eye of the Beholder" and *Gattaca* both speak to a very interesting and frustratingly complex area of inquiry referred to as "genetic-environmental covariance."

The general idea here is that societal forces will act upon people in different ways, depending on their genetic makeup. If a teacher were to believe, even unconsciously, that girls were not inherently as capable as boys in areas such as science or math, and thus was less likely to call on girls or invest the same time in helping them understand, what would the result be? Girls in that class would not fare as well as the boys.

In this vein, a recent study for the National Bureau of Economic Research (a private nonprofit organization) revealed that when exams were regraded by teachers blind to students' gender, girls' scores increased and boys' decreased, showing the presence of significant bias in the way that teachers normally assess their students.

This is a complex problem with multiple negative consequences, but a girl who unfairly doubts her innate abilities in science and math will certainly be less likely to achieve than a boy who is encouraged throughout his academic experiences. Similarly, recent work out of Yale University suggests that as early as preschool black boys can be subjected to prejudicial treatment by their teachers that may make them more likely to be suspected of perceived behavioral infractions as well as more severely punished. Thus, black boys may over time unfairly earn bad reputations and the concomitant repercussions that come with being seen as a miscreant.

Genetic-environmental covariance is cited as an alternative interpretation to many other observations that are used as evidence for inherent superiority or inferiority of one group relative to another. Richard Herrnstein and Charles Murray, in their controversial 1994 book, *The Bell Curve*, took great pains to carefully compare the results of IQ tests taken by African Americans and people of European descent and put their analyses into a societal context. The authors accounted for socioeconomic factors in their data sets and declared unequivocally that blacks are inherently less intelligent than whites. They even went on to suggest that this might lead to a harmonious future in which we give up on programs like affirmative action that attempt to shoehorn those with inherently lower IQs into positions that should be held only by the "intellectual elite." This is definitively a load of crap.

Herrnstein and Murray suggest that there exist two causes by which a person's IQ might be determined, their genes or their environment, basically the nature-or-nurture debate. The only way to understand both, they maintain, is to fix one and look at the other. So they described studies of twins who had been separated at birth and determined that two people with identical genomes that are raised in completely separate environments will often end up with similar IQs. However, when looked at more carefully, these twin studies have a large number of flaws that make them dubious at best. Regardless though, even if you could understand genetic differences in the same environment, such as a family full of foster children, or could conduct a series of better-controlled twin studies, that would still leave out a significant piece of the puzzle: genetic-environmental covariance. How could you ever quantify the effect to which genetically determined characteristics, such as sex or race, will lead to alterations in experience—in essence, environment?

The power of genetic-environmental covariance has also been ignored by a number of other highly publicized conclusions in this area. In 2007, none other than James Watson, of Watson and Crick, gave an interview in which he stated that he firmly believed blacks to be intellectually inferior to whites. Although Watson's comment was called out by the scientific community and publicly refuted in numerous venues, it is not the only dubious thing he has said since his meteoric rise to fame, as you will discover later in this book.

PART THREE

THE AMAZING
TECHNOLOGY OF
SEEING IMPOSSIBLY
SMALL THINGS

CHAPTER 12

The Jellyfish That
Taught Us How to See

I N THE FIRST CHAPTER OF THIS BOOK, YOU READ HOW AN AMAZING
new world was revealed by the invention of the microscope in
the seventeenth century. However, there was still a big problem with studying cells visually. The structures and organelles inside
the cell are pretty much all translucent and colorless. It's a bit like
looking at plastic bags filled with cloudy water in a glass bowl full of
cloudy water.

Scientists needed a way to label specific cells, proteins, subcellular
structures, and compartments to provide specificity, the ability to see
only what we want to see, and contrast, making the object of interest
easier to distinguish from its surroundings. This problem persisted
for about three hundred years after Robert Hooke first described
the cells he saw through his microscope. Stains that increase contrast, improving the ability to distinguish what you want to see from
the background, have been instrumental in clinical diagnostics and
pathology for a long time. The most famous of these is probably the
H&E stain, which is a combination of the chemicals hematoxylin
and eosin. H&E has been used for over one hundred years to stain
tissue sections and blood samples to improve the ability to visualize

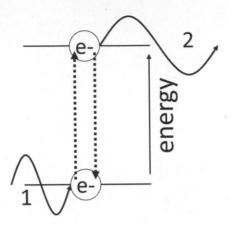

What makes a jellyfish glow? (1) A fluorophore is excited
by a photon of relatively low wavelength that increases the
energy state of an electron in the fluorophore, which then
(2) emits a photon of higher wavelength and lower energy,
returning the electron to the lower energy state.

cells and other structures such as connective tissue. However, stains
such as these do not provide molecular specificity.

In the twentieth century, scientists began using the technique of
fluorescence microscopy, in which a label that emits light of a particular color reveals detailed information about the inner structure and
workings of the cell. However, this technology was limited to two
different types of experimental applications. Researchers could add
fluorescent molecules to live cells, hoping they would end up in the
proper places, marking the compartments and structures of interest.
The process is about as imprecise as using stains such as H&E. On
the other hand, structures and components can be very precisely and
specifically labeled in dead cells. What was lacking was an easy way
to mark a specific protein in a living cell. Such a technology was
developed recently with the help of observations gained by studying
jellyfish and other sea creatures such as corals that are inherently
fluorescent and glow in the dark depths of the sea. The ability to label
proteins in a living cell with fluorescence has now become a standard

approach in research laboratories worldwide. In 2008, the Nobel Prize was awarded to the heterogeneous team of scientists whose work led to the revolutionary technology of fluorescent proteins.

To grasp how they did this, it will help you to have an understanding of what fluorescence really is, so you might need to brush up a bit on your high school physics, especially in regard to atoms and light. Electrons are negatively charged particles that form a shell around atoms. Electrons weigh very little and fly around the atom's nucleus, in which reside the positively charged protons and the uncharged neutrons. Electrons weigh about 2,000 times less than protons or neutrons and whizz around in a cloud at 2,200 km/s, or about five million miles per hour. Electrons seem to orbit around the atomic nucleus more or less as the planets in our solar system move around the sun.

Photons are the currency of light. They are how the form of electromagnetic energy called light travels. Sometimes, when a photon hits an electron, something strange happens. The electron will absorb the energy of the photon and then a short time later release a second photon. This is the basis of fluorescence. Molecules that will regularly and predictably absorb and release photons are called fluorophores. The prefix "fluor" comes from the word fluorite, a mineral form of calcium fluoride, CaF_2. Fluorite glows when ultraviolet light is shined on it, for the same reason that a white T-shirt glows under a black light, so called because it emits ultraviolet light, which the human eye cannot see. The trick is that the black light causes fluorescence in your T-shirt in the form of visible light.

The glowing white T-shirt represents a valuable insight into an important aspect of fluorescence. All light travels at the same speed, 300 million meters per second, or 670 million miles per hour, in the vacuum of space. Light travels a bit more slowly when it has to deal with matter such as air molecules. The color of light, then, doesn't depend on the speed of light, which is constant, but on two other factors, wavelength and frequency, which are related in this equation: wavelength times frequency = the speed of light.

Although photons act like particles, they can be measured and classified like waves. Waves are regular oscillations in energy that move through some kind of medium. Think of the waves that you see breaking at the beach. If you were looking down at the crashing waves from a bluff high above, you would see that they generally come in a regular series, each wave distinctly separated from the next. The wavelength is the distance between one wave and the next. Sometimes waves are short and crash in rapid succession, and other times they are far apart. The frequency, then, is how often the waves crash. The longer the wavelengths, the less often they crash, and the more often they crash, the shorter the wavelength. This is referred to as an "inverse proportionality," and it is the way light waves work. The product of the two (wavelength multiplied by frequency) does not change, as it is in fact the speed of light, which, as you know, is a constant. If one increases, the other decreases correspondingly. If wavelength doubles, the frequency decreases by half, and vice versa.

This leads us to one other facet of light that can be measured—energy. Not all photons carry the same amount of energy. Imagine that you are standing in the water just off the beach with waves breaking on you. The more waves that break in any particular period of time, the more energy is released onto you. Assuming that all the waves are the same height, if you stood there for ten minutes while the waves came at the rate ten per minute, you would have to withstand more energy than if they came in at the rate of only five per minute. It is the same with light. The higher the frequency of light, the higher the energy; energy and frequency are directly proportional. Furthermore, since we know that frequency and wavelength are inversely proportional, this means that the lower the wavelength, the greater the energy, and vice versa. Please hold on to that notion a moment while we discuss color.

Color is a function of wavelength. While the visible spectrum is a continuum of changing colors, the wavelengths for the different colors of light break out approximately like this:

Color	Wavelength
Blue	450 nm
Green	550 nm
Yellow	600 nm
Orange	650 nm
Red	700 nm

All wavelengths are approximate.

Remember the black light we discussed earlier? The reason you can't see this light is that it is actually ultraviolet light, which has a wavelength less than 400 nm, outside the spectrum that is visible to humans. The reason you can see the light emitted from the white shirt is that the wavelength of emitted light is greater than that of the ultraviolet excitation light shined onto the fluorophore. This is the key criterion for all fluorescence microscopy. It occurs because there is a very small but perceptible decrease in energy between the excitation and emission that is caused by factors such as heat loss during the process, essentially molecular friction. In short, the energy of the emitted photon is lower than the photon that was used to excite the fluorophore.

Since energy is proportional to frequency, and inversely proportional to wavelength, this loss of energy means that the emitted photon will have a higher wavelength. A fluorophore excited by invisible ultraviolet light will emit visible blue light. Similarly, we use blue excitation light to excite green fluorophores, and so on. This phenomenon is called the Stokes shift after George Stokes, the nineteenth-century scientist who studied the properties of fluorite and, in a seminal paper in 1852, described the phenomenon of fluorescence for the first time.

As previously discussed, fluorescence microscopy has long been employed, but with limited effectiveness. Scientists either had to inject fluorophores into living cells and hope they settled in the places they wanted to see, which wasn't very reliable, or to use dead cells. In

this latter approach, cells are "fixed" with chemicals such as formalde-
hyde, meaning that the cell is killed and its structures locked in place.
We then give definition and contrast to these structures by means
of fluorescently labeled antibody proteins that bind very tightly to a
specific cellular constituent.

Antibodies are proteins made by the immune system to identify
foreign pathogens, and they are useful in fluorescence microscopy.
We inject a purified sample of the target of interest, such as a specific human protein, into a lab animal—a mouse or rabbit—and wait
for the animal to mount an immune response against this foreign
invader. Then we collect the antibodies, label them with fluorophores,
and use them in fixed cells to identify the location of that target
protein. While fluorescently tagged antibodies can also be added to
live cells, it is very tricky to get this to work with adequate specificity
and contrast.

What scientists lacked was a way to easily and reliably label specific structures, components, and proteins inside living cells.

Enter the jellyfish. The species in question, *Aequorea victoria*,
grows to about 10 cm (4 inches) across and glows bluish green. The
stuff that makes it glow is creatively named green fluorescent protein
(GFP). The discovery, study, engineering, and application of GFP has
resulted in a revolution in cell biology. Now any protein of interest
can be imaged by fluorescence microscopy in live cells.

When GFP is excited by blue light, it emits green light. Also,
GFP has a very compact barrel-shaped structure, which means that it
can be linked to nearly any protein of interest without compromising
the structure, function, or cellular location of the target protein. The
way proteins of interest are labeled with GFP is ingenious, although
the methods used were actually not invented just for this purpose.

A GFP linked to a protein of interest is called a fusion protein,
or, more poetically, a chimera. A chimera is a mythical beast that is a
combination of several animals, such as a lion, a snake, and a goat, as
described by Homer in *The Iliad*. The way to make a protein a chimera

is simply to introduce into a cell a single DNA or RNA sequence that has been engineered to encode your protein of interest as well as GFP, one right after the other, with no "stop" codon between them (see chapter 3). This causes the gene-expression machinery to read the whole thing as one large protein, which, when generated, will have all of the characteristics of the target protein, but will be inherently green fluorescent. The GFP will hang off the end of your protein like the headlight on a motorcycle in your rearview mirror, letting you know exactly where it is.

Amazingly, more often than not this actually works; the GFP-fusion protein is expressed and behaves just like the unlabeled version of the protein normally found in cells. It should be mentioned, however, that the original GFP that was isolated from the jellyfish was actually not very well suited to cellular microscopy; it took a great deal of work to develop GFP into the powerful tool it is today. Comparing the natural GFP that the jellyfish normally expresses to that used in the lab is a bit like comparing a candle to a high-intensity electric lightbulb.

The process of getting GFP out of the jellyfish and into the lab involved several steps, the first taken by Osamu Shimomura, a Japanese biochemist who was interested in studying fluorescent organisms and, in particular, fluorescent jellyfish collected from the ocean off Washington State. Shimomura received a portion of the Nobel Prize for his discovery of GFP, but it was the two other corecipients, Martin Chalfie and Roger Tsien, who developed GFP into a revolutionary tool for cellular microscopy.

GFP, as it exists in the jellyfish, is not particularly bright or stable. Its fluorescence dies fairly rapidly, especially if bright light is being used to excite a small number of molecules in a living cell. This is called photobleaching. Mutants were generated that contained alterations in the DNA that encodes the GFP to display improved brightness and decreased photobleaching. Also, it was shown that GFP didn't have to be green. Mutations could be made that would change the color of GFP, so that a whole spectrum of GFP variants, such as blue (BFP) and yellow (YFP), could be produced. Interestingly though, a red fluorescent protein was never successfully generated from GFP. Apparently this is beyond the capacity of the structure of the GFP fluorophore.

Nonetheless, a red fluorophore was highly desired, since red and green are very easily differentiated in a fluorescence microscope, being relatively far apart in the color spectrum. This was a major problem. Scientists had to look elsewhere for a red fluorescent protein, and they found one, DsRed, in a coral of the genus *Discosoma*, that has been employed to great effect. The combination of fluorescent protein variants derived from GFP and DsRed has been extremely powerful when two different fusion proteins are imaged at the same time, especially in live cells.

Like GFP, DsRed was not fully adequate for cellular imaging and required some genetic engineering. Unlike GFP, the naturally occurring DsRed was quite bright and photostable. The problem was that, when linked to proteins of interest, the DsRed often caused strange things to happen. Quite often, the DsRed-fusion protein would form large clumps inside of cells, called aggregates. Thus, the protein being studied would not be in a normal, functional configuration or cellular location. This hindered many useful experiments.

Investigation into the nature of the DsRed protein turned up the cause of this problem and a potential solution. As you read in chapter 3, some proteins are made up of multiple copies of the same polypeptide, bound together. These are called multimers. DsRed normally exists as a multimer. In order to function as a red fluorophore, four copies of the DsRed polypeptide need to be stuck together. This is called a homotetramer. Unfortunately, many of the target proteins that people want to study do not respond very well to being stuck together this way. Imagine that you were running a marathon at night, and you wanted to wear glow-in-the-dark shoes. If those shoes only functioned properly when four of them were tied together, each on a different person's foot, you wouldn't get very far. So, in an effort to solve this problem, Roger Tsien generated mutant variants of DsRed that consisted of only a single copy of the polypeptide sequence, were monomers, and still glowed bright-red fluorescent. Now the combination of GFP and DsRed variants provide the ability to image processes, compartments, and pathways in multiple colors in live cells.

Fluorescent proteins have enabled an amazing variety of experiments that couldn't be performed previously.

Scientists have created mice in which every cell expresses GFP. These mice actually glow a ghostly green when viewed under bluish light. Single GFP proteins floating around inside a living cell have been imaged. The flexibility and generally innocuous nature of GFP have led to an extremely wide range of applications. Cancer cells labeled with GFP have been transplanted into a mouse to track the process of metastasis in real time. Nearly every type of protein in the cell has been studied by making GFP fusions. In fact, a collection of different yeast strains, each expressing a different single GFP-tagged protein, has been genetically engineered through this technology that allows the study of nearly every yeast protein. GFP variants have been generated that get brighter in specific environments, such as in the presence of particular ions. A series of GFP fusions that will reliably mark nearly every subcellular organelle or compartment has been identified to be used as standard cellular markers for assessing the location of other proteins. If you tag a protein of interest with a red fluorescent protein, and it localizes to the site of one of these well-known GFP-tagged markers, you now know the specific organelle where the red-tagged protein resides. Twenty years ago, none of this was possible, and now you cannot turn the pages of any cell biology journal without seeing images of GFP. All this from one jellyfish! And the fluorescent-protein story is far from over. As with the identification of DsRed in the coral *Discosoma*, other fluorescent organisms may yield the potential for new and exciting discoveries. It should be stated that, although proteins tagged with fluorescent proteins usually localize and function normally, there have been some cases where an appropriate GFP fusion cannot readily be produced. That said, the ability to genetically tag nearly all proteins of interest with a fluorescent protein has revolutionized biology by allowing scientists to use their fluorescent microscopes to image live cells in action.

How do fluorescent microscopes work? Please read on . . .

CHAPTER 13

How We See Clearly
Inside Living Cells

◉

I N THE PREVIOUS CHAPTER, YOU LEARNED ABOUT THE USE OF fluorescence in making structures and proteins inside cells more distinct and clearly visible. Now let's talk about the second word in the phrase *fluorescence microscopy*—that is, the instruments through which we look at cells.

Microscopes magnify collected light through lenses. We view these magnified images either through a microscope's eyepiece or by means of a digital collection device such as a camera. The microscopes that Hooke and Leeuwenhoek used only had the ability to collect ambient white light emitted from a candle or the sun that had passed through a sample. This is referred to as transmitted-light microscopy. What you see is basically the sample in relief, with the denser parts blocking light and the less dense areas allowing light to pass through. It's like holding up a dry leaf to the sun and seeing the internal structures revealed, except that what you see is greatly magnified. There are ways to improve the contrast in a transmitted-light image. One way is by staining the sample. For example, you might inject ink into the vascular system of a leaf to make the veins stand out, or use H&E stain as described previously.

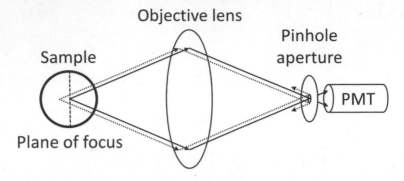

In a confocal laser-scanning microscope, light emitted
from the sample is collected by the objective lens. A
barrier with a pinhole aperture blocks out-of-focus light
(dotted lines) so that only the in-focus light (solid lines)
gets through the pinhole, allowing it to be detected by the
photomultiplier tube (PMT).

There are also optical methods for increasing contrast using special lenses, prisms, and filters. However, transmitted-light microscopy at best gives only a rough idea of the shape and overall internal
structure of your sample. It is hard to obtain a high-contrast image
of the subcellular space, as most constituents in a cell block light to
about the same degree, nor are there good ways to label all the different types of compartments, organelles, and molecules to be imaged
with transmitted light. This is where fluorescence microscopy comes
in. It provides greater contrast and higher specificity in allowing you
to see just the components that you are studying.

In the previous chapter, you read about the ingenious ways scientists now are able to insert fluorophores into cells to make components they want to see specific and distinct. We excite the fluorophore
by illumination with the particular wavelengths of light that will
cause the fluorophore to emit light in response. Excitation light is
often generated by a lamp that produces all wavelengths of light. This
illumination source is generally on the same side of the sample as the
detection apparatus, rather than behind it, as in transmitted-light
microscopy. The excitation light is passed through a filter that blocks

all wavelengths except that desired to excite the fluorophore, and then ultimately focused through the objective lens, the part of the microscope about the size and shape of a shot glass directly adjacent to the sample. The same objective lens then collects the light emitted by the fluorophores in the sample.

This configuration is referred to as epifluorescence microscopy. An instrument that has the objective lens pointing down at the sample is called an upright microscope, and one that points up at the sample is called an inverted microscope. Upright microscopes are good for imaging thicker samples and specimens that are not sandwiched between a coverslip (a very thin piece of glass) and a slide, for example, living model organisms such as fruit flies or worms in an open dish. The inverted configuration allows you to use the highest-quality objective lenses, which employ a thin layer of optically transparent oil between the lens and the coverslip to facilitate high magnification and high-resolution imaging. Passing the light rays through oils rather than air generally gives us the ability to take higher-magnification and higher-resolution images.

Epifluorescence microscopy works extremely well for many applications, but it has one major limitation, which has to do with depth of field. It's similar to when you take a photograph of a person in a crowd using a long lens—when you carefully focus on your subject, you notice that the faces of people closer to your camera are out of focus. It's the same when you use a macro lens for a close-up of some flowers, and the leaves behind them are out of focus. The depth of field is very shallow. Similarly, shining your excitation light into a cell will cause fluorescent molecules to glow within, above, and below the plane of focus. Thus, for samples that have fluorophores in more than one focal plane, which is pretty much anything biological, you will often get a blurry image.

If a cell is 10 microns thick and we use a good-quality high-magnification objective lens, the depth of field is only a few hundred nanometers. That means that everything above or below that plane will be blurry; you can only focus on a small amount of the total volume of the cell at once. This can often lead to images so blurry that you can't clearly see the structures in the focal plane you are interested in studying.

The most widely used solution to the problem of out-of-focus light is an invention so elegant and effective that it has served to define how nearly all cell-biological research is conducted in the almost thirty years since its introduction. The answer is to simply block the unwanted signal from being collected. This is accomplished by using a "confocal" microscope, which is designed so that only the light that is in focus reaches the detector. The basis for confocal microscopy is a physical barrier in the light path between the sample and the detector called a pinhole aperture. This is essentially a little hole in the light path so small that only the in-focus light emitted by the sample can pass through it. All the out-of-focus light is blocked. This technology generates incredibly sharp images of only a single focal plane.

In this ingenious design (see diagram at the beginning of this chapter), the pinhole moves up and down in concert with the plane of focus of the sample. In this way, you can focus on, and only collect light from, the top, middle, or bottom of your sample. You can even take a series of images stepping through every focal plane. Using image-processing software, these "stacks" of images can then be rendered into 3D volumes.

Although the fundamental physical concepts behind confocal microscopy have been around for nearly fifty years; the design that ultimately won out is generally credited to a scientist working in the United Kingdom in the 1980s named Brad Amos. While he did not invent the concept of confocal microscopy, and did not work alone, he was primarily responsible for the configuration that remains the preferred technology for cellular imaging worldwide: the confocal laser-scanning microscope (CLSM). It is hard to think of another technology that has so fundamentally solved a problem that limited scientific progress.

The CLSM does have one limitation. Epifluorescence microscopes illuminate the entire field of view at the same time and collect all of the emitted light in a single camera frame; this is referred to as wide-field imaging. A confocal microscope, because the pinhole can only allow a single small beam of light to pass through it, illuminates and collects an image of one tiny spot at a time. The excitation of the fluorophores is achieved by focusing a laser beam on a spot only a few hundred nanometers in diameter. A full-field image is made by scanning horizontally across the specimen, taking a series of these tiny images, one pixel each, and then dropping down and scanning a new row. This is called raster scanning. It's analogous to the way an inkjet printer works, spitting out one horizontal line of ink at a time and then moving down to the next.

Most conventional laser-scanning confocal microscopes employ photomultiplier tubes, or PMTs. These receive the light from each spot one at a time and convert it into a digital signal, sort of like a camera with only one pixel. As the laser raster scans across the field of view, the PMT records the emitted light, and the computer notes the position and intensity. Once the entire field of view has been scanned, the computer recombines all the pixels, like laying out a mosaic floor, and generates an image that renders the sample field of view *only at the plane of focus*, since all light from above and below the plane has been blocked by the barrier surrounding the pinhole. The confocal laser-scanning microscope can be applied to many different types of samples, from cells to tissues to whole organisms. It can image multiple fluorophores at the same time and generate high-resolution images and volumetric views of biological samples with incredible clarity.

And yet, the confocal microscope has several drawbacks, especially with regard to looking at living cells. It generally takes at least a second to scan the laser beam across the entire sample plane. Proteins and structures in living cells can whiz around so fast that images can end up distorted and blurry, like a picture of someone throwing a baseball taken with a nineteenth-century camera that required a very long exposure time. Further, because most of the light is rejected by the pinhole barrier, and the PMTs are not very sensitive, you have to shine a lot of excitation light onto your sample, which can rapidly cook living cells and bleach your fluorophores. Confocal laser-scanning microscopes are being developed with faster scanners and more sensitive detectors that can image live cells faster than one image per second, but as of now there are few choices for imaging live samples. A "spinning-disk" confocal microscope uses an array of pinholes placed in, you guessed it, a disk that is spinning. This allows the excitation light to raster scan across the sample very quickly, so fast that the image can be collected with a camera rather than a PMT, as if the illumination were wide field. As digital cameras are generally more sensitive than PMTs, this technology can image more quickly than a conventional confocal laser-scanning microscope, and more efficiently, using less illumination light. With a properly equipped spinning-disk microscope, tens to hundreds of frames per second can be imaged. This makes spinning-disk microscopes a good alternative to conventional confocal for imaging live cells. However, the array of pinholes on the disk is not as good at blocking out-of-focus light as the single pinhole, so you end up with images that are slightly blurry. Also, most spinning-disk microscopes can't image multiple colors at the same time.

Both confocal and spinning-disk microscopes are limited in how deeply they can see into samples, so they are best used for imaging individual cells. This is because the objective lenses used for the highest resolution and magnification do not have very long "working distances," the maximum distance allowed between the lens and the sample imaging plane. For the highest-quality objective lenses, this is only about 100 microns. That is fine for imaging cells that are 50 microns tall, but what if you want to look into tissue or a live model organism? Another problem is that visible light scatters easily, so that even if you could use an objective lens with an extra-long working distance, you would not be able to see very much—the excitation light would never reach the fluorophores deep in the sample. It's like trying to shine a flashlight through your hand. Even with a very bright light, you can only see through the thinnest and least dense parts. Infrared light, which has wavelengths too high to be visible, doesn't scatter much, so that it penetrates very efficiently through tissue. However, the fluorophores we use all emit their light in the visible range.

Two-photon confocal microscopes make use of a fantastic phenomenon that many fluorophores demonstrate: if two photons of approximately double the excitation wavelength simultaneously strike the fluorophore, they will effectively combine into illumination as if a single photon were encountered at the normal excitation wavelength. GFP is normally excited with a 488 nm wavelength light. However, in a two-photon microscope, the best wavelength for excitation is somewhere around 970 nm. That's well into the infrared range, which solves the scattering problem while causing the fluorophore to emit visible blue-green light at about half that wavelength.

A further benefit of two-photon microscopy is that the odds of two photons striking the fluorophore at the same time are very low. This means that when focusing a laser at the structures you are interested in, fluorophores outside of this focus plane will not be excited. Because simultaneous absorption of two photons is a low-probability event, it only occurs at the plane of focus. This means that a pinhole in the light path is not needed, since no illumination light is exciting fluorophores above or below the focal plane. However, the infrared lasers used for two-photon microscopes are extremely expensive and, as the long working-distance lenses don't provide optimal resolution or magnification, this technique is often reserved for imaging thick tissues and model organisms, rather than cells.

There is another alternative to confocal microscopy that is fast and useful for live-cell imaging—basically epifluorescence microscopy. You don't need expensive lasers or pinholes—just a lamp, an objective lens, and some filters. Furthermore, you don't throw away any of the light so you can get images with very low excitation intensities—you don't kill the live cells you want to look at. This efficiency also can allow you to image very quickly with low exposure times on the digital camera, great for making movies of quickly moving structures.

The only problem is that pesky blurring caused by the fluorophores outside the focal plane. The solution is the use of computers, or more precisely, math.

Various factors, such as the depth of the focal plane of the wavelength of light being emitted and the magnification of the objective lens, affect the particular type of haze that will be formed by each fluorophore. With this knowledge, you can use a mathematical framework to computationally generate focused images from blurry ones. This is referred to as "deconvolution," and to perform it you need an understanding of the "point spread function." This has nothing to do with sports betting. It refers to the haze that spreads in three dimensions from a fluorophore. You can estimate point spread function if you know enough about your microscope. An even better way is to determine it empirically by imaging small structures of known dimensions, such as small fluorescent-plastic beads, comparing the beads that are in focus to images of beads at precisely known distances above and below the plane of focus. Once you determine the point spread function for your particular microscope, deconvolution software can be used to take the out-of-focus haze from images after they have been acquired and even reassign the light to the correct focal plane. Thus, in addition to providing sharp individual images, this technology can also be used to create a volume view by acquiring a stack of epifluorescence images, as in confocal microscopy.

While epifluorescence with deconvolution can be very fast and generate high-quality images, it can be complicated to set up the system so that the software's ability to reassign the light to the correct focal plane is truly precise and reliable. Moreover, with all of these techniques, you are still shining light above and below the plane of focus, which means you are unnecessarily damaging your sample with out-of-focus light. So why not just shine the light only where you want it?

That's our next chapter.

CHAPTER 14

Light-Sheet Microscopy, or, The Light in SPIM Stays Mainly in the Plane

◎

D ESPITE ALL THE ADVANCES IN FLUORESCENCE MICROSCOPY that we've discussed so far, a couple of fundamental problems remain unsolved.

In conventional forms of fluorescence microscopy, the illumination light excites fluorophores throughout the volume of the sample, including outside of the plane of focus. Each of the methods we have previously discussed that can solve the problem of blurry images is either slow, delivers too much high-intensity light to keep cells and organisms from dying, or can't travel very deep into living tissue. One key issue is that these techniques all employ a single objective lens for both illumination of the sample and collection of emitted light.

The next brilliant idea in microscope design began with this question: What if you turned the conventional microscope on its ear? Microscope systems are now being produced with two different objective lenses, one for the excitation light and a second for collecting the emission. This configuration is the basis for what is called selective plane illumination microscopy (SPIM), or light-sheet microscopy. The technique uses special optics in the excitation light

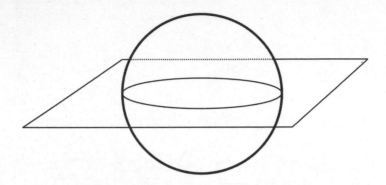

Selective plane illumination microscopy (SPIM), also
called light-sheet microscopy, involves creation of a
horizontal plane of illumination that excites fluorophores
only at the plane of focus, for greater image clarity.

path that make the illumination into a flat plane of light. The idea
is that if you illuminate your sample with a flat horizontal plane of
light, rather than exciting fluorophores above and below your plane
of focus, you can shine the light where you want it. To image through
the volume of your sample, all you have to do is raise or lower the
light sheet as you focus up and down.

Light-sheet microscopy has several benefits. Since the light
sheet results in wide-field illumination—shining light across the
entire horizontal sample plane—you can use a very fast and sensitive
camera for imaging. Furthermore, the light sheet can move up and
down very quickly, so you can make image stacks through even large
samples very quickly. However, the most critical benefit of SPIM is
that no light shines on the sample outside of your focal plane. This
means that the rest of your sample avoids the damaging effects of
high-power illumination. Although SPIM has only been commercially available for a few years, it already has been adopted by a wide
range of researchers. SPIM can be used to image each and every
cell in a developing embryo—for example, from a *Drosophila* fruit
fly—during early rounds of cell division and differentiation. SPIM

can also be used by cell biologists to image individual living cells for hours or even days.

SPIM also has drawbacks, though, that make it inapplicable for some types of studies. In particular, the geometry of these systems is such that the same microscope can't be used to image both small individual cells and larger model organisms. The optics and sample holders are just not flexible enough. Further, for thicker samples like embryos, the same limitations exist as with two-photon microscopy; extremely high magnification and spatial resolution are difficult to achieve. This is because you need an objective lens with a longer working distance to be able to focus through a thick specimen, and these don't have the same degree of magnification and resolution as lenses used for thin samples such as cells. However, the most critical drawback of SPIM lies in the issue of scattering. We've discussed how light scatters when it hits biological tissue. With a SPIM system, the excitation light illuminates the sample from one side, and the emitted light is collected through a lens perpendicular to this plane. This means that if you image a large sphere, only the portion of the sample closest to the excitation source and the collection lens will be effectively imaged—only about a quarter of the sphere. This scatter is a real problem, though some newer SPIM systems give you the ability to rotate the sample so that each side can be imaged at the highest level of efficiency. Others have optics that allow two different light sheets to be generated, one from either side. In this way, scatter of the excitation light can be balanced out. SPIM is not yet a fully mature technology, and new configurations are being developed all the time.

Yet another selective-illumination system has existed for decades that is extremely well suited to imaging live cells at very high magnification and resolution at very high speeds with minimal photobleaching or photodamage to the sample. However, this technology only allows you to image the very bottom of a cell that is stuck to a glass surface.

Here's how it works. If you shoot a laser beam at a piece of glass at a ninety-degree angle perpendicular to the surface, it will go right through. However, if you shine the laser at the glass at a higher angle, the light will seem to bend. This is refraction, and it is the same phenomenon that makes a pencil in a glass of water look bent—light travels more slowly in dense substances. If you keep decreasing the angle at which the laser strikes the glass, at a certain point—the critical angle—the light will no longer enter the glass at all but will reflect off of it. This is called total internal reflection (TIR), and it is the reason expertly cut diamonds appear to glow.

When TIR occurs, the surface of the glass glows in what is called an evanescent wave. The evanescent wave is easily overlooked, as it only penetrates a very short distance beyond the surface of the glass. However, total internal reflection fluorescence (TIRF) microscopy, also known as evanescent-wave microscopy, harnesses this glow to amazing effect. If you shine a laser through your objective lens in such a way that it strikes the bottom of a coverslip at an angle greater than the critical, you will generate an evanescent wave that can be used for fluorescence imaging. The evanescent wave will only penetrate about 100 nm into the sample, so TIRF can only be used to image the very bottom of a cell that is stuck to a coverslip. However, a great deal happens at a cell's plasma membrane, such as cell adhesion and receptor signaling, and TIRF microscopy is excellently suited to these types of studies, especially in living cells. That is because no light penetrates into the cell beyond this very thin plane emanating from the coverslip. TIRF microscopy spares the majority of the cell from the damaging effects of high-energy laser light while providing crisp, sharp, high-contrast images. Also, a sensitive camera can collect TIRF images extremely rapidly. The one big drawback is not being able to see deeper into the cell.

So today we have all these technologies that solve, in varying degrees and applications, the problems of seeing things clearly, in focus, with good contrast, and without killing live cells. Yet one big obstacle remains: resolution. There is only so much detail we can see by focusing light through glass lenses. In particular, the lenses that are used for two-photon microscopes or SPIM that let you image deeper into samples do not permit very high resolution. Even with the highest-magnification oil-immersion objective lenses used for confocal or TIRF microscopy, employed for optimum focus, contrast, and clarity, we are still unable to see things inside the cell as they really are. Beyond that point, it's like trying to play piano while wearing mittens. However, a series of remarkable innovations in super-resolution microscopy is now revolutionizing the way we see inside of cells.

CHAPTER 15

Super-Resolution Microscopy: Turning the Lights On One at a Time

◉

F OR ALL THE ADVANCES WE'VE SEEN IN FLUORESCENCE microscopy—technologies that maximize contrast, clarity, and sharpness of the image—there is still a limit to resolution using glass lenses.

The structures, molecules, and organelles inside of cells are very small and often very close together. The synaptic vesicles that carry neurotransmitters might be only 50 nm across and can cluster tightly together. With a conventional microscope, even a confocal, you will not be able to distinguish individual vesicles. The image would look like a silhouette of a bunch of grapes. The fibers that make up a cell's skeleton, the cytoskeleton, can bundle together like a bunch of dry spaghetti that you hold in your hand just before dropping it into the pot. If that pasta glowed and you turned out the lights and took a picture, you would not be able to resolve each individual piece. It would look like you were holding a single glowing rod or cylinder. That is the problem with the resolution limit of light microscopy. Objects that are closer together than about 250 nm will merge together in your image and not look like individual, separate structures.

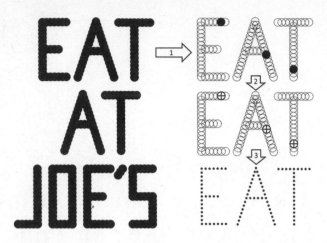

In single-molecule localization microscopy techniques, (1) only a few fluorophores blink per camera frame. (2) By taking thousands of frames, the precise location of each blinking event can be recorded so that (3) an image is reconstructed at molecular-scale resolution.

We have all seen old-fashioned lighted signs that are made up of many lightbulbs spelling out the words, the kind used on theater marquees and roadside diners. When you first spot that EAT AT JOE's sign from a couple blocks away, you can't see the individual lightbulbs, that is, you can't resolve them. This is exactly what happens with fluorescence microscopy using a normal microscope. The image you get is of an ensemble of fluorophores, the light from which merges to give an overall impression of a shape or structure. Increased magnification can only do so much to solve this problem, as the issue is really the overlapping of light waves emanating from the closely adjacent fluorophores. When the groups of photons emitted from individual adjacent fluorophores are too close together, they cannot be distinguished from one another, and the individual fluorescent structures cannot be resolved. So then how can you possibly see each lightbulb in the EAT AT JOE's sign?

Recently, several techniques have been developed for improving the resolution of the fluorescence microscope beyond the conventional limit, collectively referred to as super-resolution microscopy. Some employ sophisticated microscope designs or complicated computational algorithms, while others rely on tricks that you can play with fluorophores.

The simplest and most elegant solution is also the best in that it results in the highest resolution. The concept is simple: make fluorophores blink—find a way to switch them on and off.

Think of it this way. Your job is to count the number of lightbulbs in the Eat At Joe's sign. When they're all turned on, you can't resolve them. But if you could turn them on one at a time, then it would be easy to count them and to map their positions. By doing so, you could ultimately combine all your mapped lightbulbs into a single picture that told you everything you needed to know at the highest resolution possible.

That is the basic idea for this particular super-resolution technology. If you had a bunch of fluorophores very close to one another, you wouldn't be able to see each individual one, *unless* they only lit up one at a time. Even if you had a million fluorophores within a very small space, if only one lit up at a time, you could eventually map exactly where each one is, as long as the pixel size of your camera was small enough—with the best lenses and cameras we can localize fluorophores with a precision of about 20 nm.

This technique was developed independently and almost simultaneously in two laboratories: those of Xiaowei Zhuang at Harvard and Jennifer Lippincott-Schwartz at the National Institutes of Health. While Zhuang employed small synthetic chemical fluorophores, rather than fluorescent proteins, Lippincott-Schwartz utilized an ingenious variant of GFP she had earlier developed that could be switched on and off like a lamp. Zhuang called her technique stochastic optical reconstruction microscopy (STORM). Lippincott-Schwartz named hers photoactivated localization microscopy (PALM), because it employed her photoactivatable GFP (PA-GFP). In many other ways, STORM and PALM are much the same, and the microscopes and computer programs developed and marketed for one can generally be used for the other. In either case, the basis of the microscopy technique is to label your cells with a fluorophore that will blink, either tagged to an antibody or tagged with a fusion protein to PA-GFP or one of the similar fluorescent proteins that can be triggered to blink. The trick is to set up the system so that you will only see a few fluorophores out of the millions in your sample in any one photograph you take. Using a very fast and sensitive camera, you just sit back and record a large number of frames with a small number of blinks per frame. One manufacturer suggests that something like 20 blinks per frame over 20,000 frames is required to generate a suitable super-resolution image. With a very sensitive camera this can be accomplished in about fifteen minutes.

What makes PA-GFP blink? Before it is photoactivated, it is essentially nonfluorescent. A quick pulse of low-intensity ultraviolet light kicks a few PA-GFP molecules into the "on" state. These active fluorescent proteins will then emit green light when excited by the blue light normally used to image conventional GFP, allowing us to map them. Importantly, the high laser powers used to excite these few activated PA-GFPs cause them to rapidly photobleach, so that they can no longer emit photons. This wipes the slate clean so that we'll be seeing a different set of blinking fluorophores in each new image— like seeing just a few distinct lightbulbs in the EAT AT JOE's sign.

On the other hand, Zhuang's STORM technique employs types of chemical fluorophores that blink all by themselves, without any cycles of illumination with different colors of light. This makes the process slightly simpler, but requires labeling techniques that are unnecessary in PALM.

Once the images have been acquired, they undergo a series of image-processing steps that identify the center point of each blinking event. PALM and STORM are both referred to as single-molecule localization microscopy, which means that the end result is not really an "image" in the way we normally consider microscopy data. Rather, these techniques provide more of a localization map, sort of like pins stuck into a chart on the wall. The pins only show where things are, not how big, small, or bright they are. The trick is knowing exactly where to stick the pins. It was actually the people who laid the groundwork for PALM and STORM, for understanding fluorophore blinking and how to image and identify the locations of individual fluorophores, who won the Nobel Prize in 2014 for "super-resolved fluorescence microscopy," namely Eric Betzig of the Howard Hughes Medical Institute and William E. Moerner of Stanford University.

Betzig and Moerner shared the prize with German scientist Stefan Hell, who independently invented an alternative super-resolution technique called stimulated emission depletion (STED) microscopy. STED is based on standard confocal laser-scanning microscopy. The idea is to suppress fluorophore emission by means of high-power light that is of a higher wavelength than that of the emitted photons. This phenomenon of emission depletion is basically like a game of molecular whack-a-mole. You hammer down the fluorophores you don't want to see so that only those of interest are revealed. Hell designed his STED microscope with two lasers focused together to scan across the sample in a conventional confocal configuration. One laser was at the wavelength for exciting his fluorophore, and the other was at the wavelength that would deplete it, preventing it from emitting light.

The truly ingenious aspect of Hell's system was that in the light path of the depletion beam he placed an optical element that would block the depletion laser, but not the excitation laser, from illuminating the center of the spot being raster scanned across the sample. The resultant donut-shaped depletion spot would scan along with the excitation laser beam, creating confocal images where only the very middle of the donut would actually reveal active fluorophores. As STED only shows you the molecules inside the donut, this decrease of the illumination spot size improves the resolution of the microscope. These benefits aside, STED suffers from the same speed limitations as conventional confocal microscopy, plus it delivers even more light to the sample. Thus, it is also not very easily performed with live cells.

Super-resolution microscopy can increase resolution up to 10 times the maximum for conventional microscopes, from about 250 nm to as small as 20 nm. Considering that this is the exact scale at which many cellular components exist, these techniques have been revolutionary. In a very few years, super-resolution microscopes have become both widely available and indispensable to researchers.

Unfortunately, neither the single-molecule techniques of PALM/ STORM nor confocal-based STED microscopy can readily be applied to live cells, which are easily damaged by intense illumination and have rapidly moving structures. Moreover, in either of these techniques, it is difficult to image fluorophores of different colors at the same time in the same sample. Although STED microscopes have been developed with multiple excitation and depletion beams, it takes a high degree of experimental planning to make sure these will all work together for effective multicolor imaging. Similarly, getting two or more fluorophores to blink in PALM or STORM can be very challenging. A good option is to combine PALM and STORM to image two different markers in the same sample, one tagged with a fluorescent protein and the other with a chemical fluorophore on an antibody.

With super-resolution microscopy, we can now discern the localization of numerous small structures that are very close to one another, such as those synaptic vesicles clustered together like a bunch of grapes. This is a very rapidly changing field. Every month there are papers reporting new fluorophores or new software solutions to simplify and expand these types of imaging studies. Super-resolution microscopes are even being developed with the potential for live-cell imaging. One such is called structured illumination microscopy (SIM). Expected to be fast and gentle enough for live cells, SIM only improves resolution about twofold, not ten times, but that would still be a significant advance. However, to this date there is still no commercially available super-resolution microscope system that eliminates out-of-focus light *and* provides the ability to image multiple colors in live cells.

CHAPTER 16

What Makes the Glowworm Glow? The Advantages of Luminescent Imaging

⊚

I N THE COURSE OF THIS BOOK, WE'VE LOOKED AT SEVERAL ANI-
mals that can emit light. The jellyfish *Aequorea victoria* glows
green because of green fluorescent protein (GFP), and the
coral *Discosoma* possesses the fluorescent protein DsRed. Fluorophores such as GFP and DsRed only emit light, that is, photons, when they are excited by incident light. There are, however,
molecules that emit light without external excitation. Rather than
fluorescence, this is referred to as luminescence. Fluorescence is a
bit like a reflector on the front of a bicycle, only glowing when
light shines on it. Luminescence is more like a lightbulb, converting
stored energy into light. Luminescent molecules are able to convert
chemical energy into light with no illumination needed. This is how
the firefly glows.

(This is not to say that the fireflies themselves are not *excited*.
They do in fact glow when they are interested in mating. It is probably fortunate that this kind of bioluminescent signaling has not been
selected for in humans. It may also be worth noting that a glowworm
is, strictly speaking, a firefly larva.)

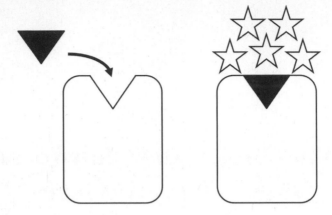

To generate light, add luciferin to the enzyme luciferase
in the presence of oxygen and ATP. Repeat as often as you
like—no excitation light needed.

Fireflies contain a protein called luciferase, which as you might
guess from its name, generates light. Fireflies are not the only animals
that display luminescence. The sea pansy, *Renilla reniformis*, which
looks like a slug a few inches long attached to a flat fan shape that is
covered in small anemone-like extensions used for filter feeding, is
another bioluminescent organism.

How does bioluminescence work? In addition to luciferase, there
is another molecule floating around inside fireflies required for the
chemical reaction that creates light. It's called luciferin, and it is not a
protein. Luciferin contains a few sulfur and nitrogen atoms arranged
in carbon rings, not unlike nucleotide molecules. In the parlance of
the enzymologist, luciferin is the substrate for the chemical reaction,
the wood destined for the luciferase saw. When the two molecules
come in close proximity, luciferase alters the structure of luciferin in
such a way that energy is released in the form of light, creating the
familiar glow of a firefly. Also required for this process are ATP, the
units of cellular energy described in chapter 2, and oxygen. The process produces carbon dioxide.

As we learned earlier, nucleotides include a chemical group called a phosphate, which consists of one phosphorus and three oxygen atoms. Often, two or three phosphate groups link together in a chain. The bonds between these phosphate links hold a great deal of energy. In fact, so much energy can be stored this way that phosphate bonds are a universal currency of energy in organisms from bacteria to humans, particularly in the molecule known as ATP. ATP is the nucleotide adenine with three phosphate groups attached to it in series—**a**denine **tri**phosphate. Breaking the bond between the second and third phosphate groups releases a great deal of energy, on a molecular scale. All life on this planet has evolved to employ ATP as the main (but not only) storage vessel for chemical energy. In the case of bioluminescence, ATP is a bit like the battery that powers the flashlight.

The ability to generate light inside cells, through chemical means and without excitation light, has several benefits for scientific research. First, when GFP or any other fluorophore becomes photobleached by the excitation light, you can't continue to image that cell or tissue sample. However, when the light from the luciferase-luciferin system runs out, you just add more luciferin. Further, sometimes you don't want any light until a particular time, for example, when a certain developmental stage has been attained. Most fluorophores will emit light when excited whether you want them to or not. But in an experiment using luciferase, you can just wait until the right moment to switch on the lights by adding the luciferin.

Also, luciferase is often used as a way to measure gene expression, for example, the level at which a particular gene is expressed under certain conditions. As discussed in chapter 5, gene expression can be regulated at the level of transcription, the process that makes RNA from DNA. Transcription of a gene can be increased or decreased depending on factors like the number of acetyl groups linked to the histone proteins around which DNA winds, or the amount of DNA methylation. A promoter is a sequence of DNA that occurs just before the coding sequence of a gene. This is where proteins known as transcription factors can bind and increase or decrease expression of the adjacent gene. One way transcription factors regulate gene expression is by adding acetyl groups to nearby histone proteins, unwinding the nearby DNA and allowing the RNA polymerase access to begin transcription. If you genetically engineer cells to express luciferase in tandem with a specific promoter from a particular gene of interest and maintain them in the presence of a constant supply of luciferin, you can use this "luciferase assay" to measure the regulatory function of that promoter. Anything that would normally affect the expression level of the gene of interest— for example, function of a particular transcription factor or activation of the cell by some specific growth factor—will now alter the expression level of luciferase and change how much light the cells produce. Luciferase assays have been instrumental in uncovering the various mechanisms that cells use to regulate gene expression.

Another extremely useful application of luciferase is in the imaging of cells within living model organisms—lab animals. Once again this is due to the fact that we can look at luminescent markers again and again, but if we are not very careful to avoid photobleaching, we may only be able to image fluorescent ones once. There are many applications in which researchers need to look at cells in an intact, living animal, for example, in testing cancer therapies. Studying the same set of animals over the entire course of an experiment is statistically more sound than testing different animals at different stages, as it minimizes the differences between individual subjects. For example, if you wanted to know whether a certain treatment reduced the size of a tumor over a period of months, you would get more reliable data by looking at the same animals month after month than if you had to look at a few animals the first month, a few others the next month, and so on. In the latter case, you wouldn't know the starting size of the tumor in each animal, so you couldn't be as sure about the effects of the treatment.

Also, by following the same animals from the beginning of the test to the end—this is called a "longitudinal" study—you can use fewer animals in your experiments. The "three Rs" of ethical animal research are *replacement*, using means other than lab animals whenever possible; *reduction*, doing everything possible to use fewer animals; and *refinement*, using techniques that minimize the discomfort to which model organisms are exposed. Longitudinal studies achieve the goal of reduction.

There is one other very significant technical benefit of using luminescent means such as luciferase rather than fluorescent proteins such as GFP, which require excitation light. If GFP is used to label cells, as in a cancer-model animal, the photons required to induce the fluorescence emission have to travel into the animal to the fluorophores. Remember our analogy of the bicycle reflector versus the lightbulb? Imagine that you had to put a thin cover over the reflector on the front of your bicycle that partially blocked the light shining onto it. It would be very dangerous to go out in traffic at night. However, if you put the same thin cover over an electric headlamp, it might still function well enough for you to see and be seen. It is the same with looking for light emitted from cells inside a living animal. Luminescence can make it much easier to see the cells you want to study. Your experiments will be much more efficient, lifting the veil to uncover new diagnostics, treatments, and therapies.

CHAPTER 17

More Ways to Take Pretty and Enlightening Pictures

T HE DISCOVERY AND DEVELOPMENT OF FLUORESCENT PRO-
teins provided a revolutionary method for imaging specific
proteins in live cells. No longer did samples need to be fixed
in place with chemicals and labeled with antibodies and fluorescent
dyes. This is important, since in many cases, ranging from the packaging of chromosomal DNA to the organization of the cytoskeleton,
chemical fixation actually forces the molecules being imaged into
unnatural structural configurations. Fluorescent proteins enabled
the development of innovative experimental procedures that deepen
our understanding by letting us watch molecular processes unfold in
live cells in real time. The great challenge that has arisen, though, is
how to keep the cells alive during imaging, as cells don't like having
high-powered laser light shined on them for a long time.

In response to that challenge, several ingenious technologies have
been invented that now help us see what goes on inside living cells in
different ways and for different purposes.

Studying the motion of specific populations of proteins within
living cells with minimal background staining has provided insights
into all types of dynamic processes. However, quantification of protein
motion can still be a challenge. There is a technique that can be employed
simply with a GFP-fusion protein that can very elegantly allow

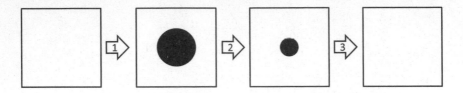

Fluorescence recovery after photobleaching (FRAP)
involves (1) deliberately photobleaching a specific region
of the cell, then (2) imaging the movement of unbleached
fluorescent molecules into the bleached region, until (3)
the bleached region is once again fully fluorescent.

researchers to study protein motion throughout a compartment, or
even the entire cells. In previous chapters, we've described photobleaching as a problem—once you've exhausted the potential of a
particular fluorescent molecule, it generally cannot be resurrected.
However, there is a process that turns photobleaching to our advantage:
fluorescence recovery after photobleaching (FRAP). The idea is that
after high-power illumination has bleached out the fluorophore markers, we measure the increase in fluorescence that happens when other
actively labeled proteins are transported back into that region of the cell.

Imagine that you wanted to study how fast the population of riders on a bus turns over as the bus travels along its route. You can only
observe the bus from a distance, making it difficult to keep track of
individual riders. But let's say all bus riders are fluorescent when they
get on the bus. You start by measuring the brightness of the light that
emanates from the windows of a full bus. Then you photobleach all
the riders at once so the inside of the bus goes dark. Then you measure
the rate at which the bus windows begin to light up again as darkened
passengers get off at each stop and new, glowing passengers get on. By
measuring how long it takes for the bus windows to glow as brightly
as they did at the start of your experiment, you'll know how long it
takes for the population of bus riders to completely turn over.

Developed in the lab of Watt Webb at Cornell University in the
mid-1970s, FRAP was initially used with fluorophores linked to

markers that would monitor the location of specific receptors on the surface of living cells. These receptors are responsible for regulating signaling pathways that control everything from cell adhesion to cell proliferation and even cell death, and the mobility of receptors in the plasma membrane often changes over time. Proteins in the plasma membrane surrounding a cell move around like leaves on the surface of a pond on a breezy day. If we labeled the proteins within the membrane, the whole surface of the cell would be fluorescent, like a pond covered with leaves. The FRAP techniques developed by Webb and his group involved bleaching small spots and then measuring the diffusion-based motion of the receptors in the plasma membrane as the bleached spots were refilled with fluorescent-labeled receptors entering from adjacent regions. This would be like removing some of the leaves from the surface of the pond and seeing how long it takes the wind to blow new leaves into the now-available space. However, this was an indirect measure, and the fluorophores had to be linked to molecules that would in turn bind to the receptor of interest. The advent of GFP led to a huge boost in the ability to perform FRAP studies, as now the proteins of interest could be directly labeled without any further manipulation.

Where FRAP lets you turn off the fluorescence of a specific population of tagged proteins, yet another technique lets us turn on fluorophores, specifically those on fusion proteins. As described in chapter 15, photoactivatable GFP (PA-GFP), which came out of the lab of Jennifer Lippincott-Schwartz at the National Institutes of Health, can be used to mark the sites of individual protein molecules using the PALM super-resolution technique. However, in live cells, PA-GFP can be used for much more than this. If a research team is interested in the movement of a protein from one compartment to another, all they have to do is photoactivate the PA-GFP at one location and then watch as it moves to another place. This reveals only the population of interest, eliminating the background staining present from previous rounds of transport.

Say, for example, you want to study the proteins that bind newly synthesized RNA in the nucleus and watch the RNA-protein complex exit the nucleus on the way to the ribosome, where the RNA will be translated. All you have to do is tag the RNA binding protein with PA-GFP and then photoactivate it in the nucleus. After the nucleus becomes green, you can track the movement of the now-fluorescent proteins bound to the RNA.

However, there is still one significant technical issue: How do you see the location you want to target with photoactivation if the PA-GFP is nonfluorescent before you stimulate it? You need a second, independent marker for the compartment within which you want to photoactivate.

An alternative class of fluorescent proteins exists that solves this problem with extreme efficiency. Photoconvertible fluorescent proteins can be transitioned from one color to another by illuminating them with ultraviolet light. For example, a fluorescent protein poetically called *kaede* (the Japanese word for maple) displays green-to-red photoconversion. In this way, you can visualize your protein of interest throughout the entirety of the cell and then photoconvert only those proteins present in the particular structure or compartment you want to study, or convert them only at a critical time. Photoconvertible fluorescent proteins can also be employed for super-resolution single-molecule localization microscopy (see chapter 15), using green-to-red transitions as markers to locate individual fluorophore molecules, instead of blinking events. Photoconvertible fluorescent proteins facilitate these types of analyses without the need of a second counterstain to label a specific subregion of the cell.

Another technique, fluorescence resonance energy transfer (FRET), was initially performed with chemical fluorophores and then evolved into a more direct approach using fluorescent proteins. Initially developed in 1946 by German scientist Theodor Förster, FRET is used to measure the distance between two proteins in a cell, specifically to determine if two proteins bind to each other. FRET can be performed in live cells and can be used to measure repeated cycles of protein binding and release in real time. It works this way: Imagine that protein A is linked to GFP and protein B is linked to DsRed.

As usual, blue excitation light is used to illuminate GFP, which emits green light. If the DsRed is close enough to the GFP, less than about 10 nm, energy from the excited GFP will cause the DsRed to light up. Importantly, it is not the green light from GFP that excites the DsRed; rather, it is a direct energy transfer, not based on photons. This is sort of like the way a tuning fork held next to a guitar string can make the string vibrate without anyone having to pluck it. Of course, it's protein A and protein B that are binding together directly, not the GFP and DsRed. The point is that the fluorescent protein tags are close enough together to allow the efficient transfer of energy from the GFP to the DsRed, telling us they are within that 10 nm range.

There is yet another way to see if two proteins bind together that uses, rather than two fluorescent proteins, only a single one cut in half. The technique of bimolecular fluorescence complementation (BiFC) employs expression of two different fusion proteins. Protein A is linked to one half of the fluorescent protein and protein B to the other half. If protein A and protein B bind and bring together the two halves of the fluorescent protein, you get fluorescence. It's like a molecular glow stick, in that it doesn't release light until you mix together the two components by breaking the barrier separating them. BiFC is simpler than FRET, as you don't have to worry about different wavelengths. Everything can be measured in a single image.

Some neuroscientists have recently made extremely good use of BiFC to determine which neurons in the central nervous system form synaptic contacts between each other. A synapse is the space between two nerve cells (neurons) across which exchange of chemical signals occurs. If you tag the outside of neuron A with one half of a fluorescent protein and the outside of neuron B with the other half, then where those two neurons form a synapse you will see a bright fluorescent spot. Known as GRASP, this technique can even be used in the brains of live model organisms like fruit flies to highlight networks of synapses.

GCaMP is a technique being applied very effectively in the study of brains in live model organisms. This is a GFP-based system that increases in fluorescence whenever a neuron is activated, letting us see beautiful lightning bolts throughout the nervous system. For example, if GCaMP is expressed in the visual cortex, that section of the brain lights up when the animal is shown visual stimuli. GCaMP works by increasing the fluorescence of a mutant variant of GFP that illuminates when the concentration of calcium in the neuron increases. When a neuron is activated, calcium ions rush into it and then are rapidly removed. We can see and record these transient spikes in calcium via the flashes of fluorescence emitted from the GCaMP.

GCaMP was not the first calcium sensor developed. Starting as a postdoctoral fellow in the 1980s, Roger Tsien, who won the Nobel Prize for his work on GFP, developed some of the first fluorescent chemical probes for calcium. Many of these are still used today. Sensors for ions other than calcium have been developed both as chemical fluorophores and as fluorescent proteins. The GFP mutant pHlourin is very sensitive to pH, and so is used in many studies to visualize the acidification of cellular compartments.

More recently, an entire new field has emerged, optogenetics, which uses proteins such as enzymes or ion transporters that can actually be activated by light. A protein is engineered so that it is nonfunctional until illuminated with ultraviolet light. This process can be repeated over time, since the engineered protein can go back to its normal, inactive state when you stop shining ultraviolet light on it. With this technology, researchers are developing tools to stimulate individual cells and track the responses in real time. As with GRASP and GCaMP, optogenetics is used not only in cells in culture, but in the central nervous systems of living model organisms.

This is a very exciting transitional time, in which researchers are harnessing the power of light to perform measurements and manipulations in live cells in truly quantitative ways, allowing us to routinely perform experiments that were not even dreamed of only a few years ago.

PART FOUR

MATTERS OF
LIFE AND DEATH

CHAPTER 18

How Cells Die

Y OU HIT YOUR THUMB WITH A HAMMER. WHILE YOU ARE howling in pain and exercising some of the more colorful portions of your vocabulary, something more serious is happening at the microscopic level inside your wounded digit. Cells are dying. In cases of externally imposed trauma, there's no way around it.

Burning your hand on a hot stove or slicing the tip of your finger with a knife sends the alarm to your immune system to rush to the scene and try to ameliorate the disaster, like the first responders rushing to a fire. But dead is dead. Inflammation and medication can prevent further damage or infection, but necrosis, the uncontrolled externally caused death of cells, will occur, and it can be dangerous. Cells contain harmful stuff. The lysosome is full of acids and digestive enzymes. Imagine pipes and earthworks in a city's waste-treatment plant being ruptured, releasing toxins into nearby rivers and groundwater. This is one of the reasons that in the time of cell damage and death, macrophage cells in your immune system are recruited to gobble up the dead and dying. (Macrophage means, more or less literally, "big eater.") Like Wile E. Coyote from the Road Runner cartoons eating a stick of dynamite disguised as a hot dog, macrophages are able to engulf dead and dying cells without suffering any permanent ill effects. Furthermore, in order to prevent the ticking time bomb of necrosis, cells themselves have a more dignified way of dying that

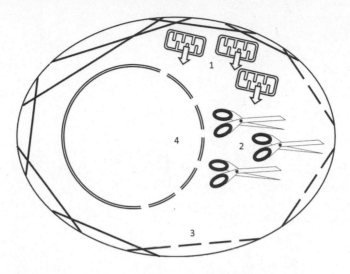

Following an increase in mitochondrial membrane permeability, (1) the contents of the mitochondria leak into the cytosol. Digestive caspase enzymes become active (2) and break down cellular components such as the actin cytoskeleton (3) and the nuclear envelope (4).

is less dangerous to their neighbors. This programmed cell death is referred to as apoptosis.

Of course, not all potently fatal insults immediately result in cell death. Cells that are starved of nutrients will begin to digest their own internal structures in a type of self-recycling. In fact, just as this book was being completed, Yoshinori Ohsumi from the Tokyo Institute of Technology in Japan was awarded a Nobel Prize for his seminar work in the study of autophagy, "self-eating." Although the mechanisms controlling autophagy have yet to be completely understood, it appears that aberrations in this process can underlie aspects of aging as well as neurodegenerative diseases and certain types of cancer.

However, once protective and preventative measures have failed, the cell enters into the process of apoptosis. There are two main types of apoptosis, but in each case, the cellular mechanisms are quite similar—it's the initial triggers that differ. In the first pathway, intrinsic apoptosis, the cell is responding to something amiss internally. Cells have mechanisms for sensing and responding to irreparable damage. The second pathway, extrinsic apoptosis, results when something external to the cell makes it preferable to burn the whole place to the ground lest the trouble spread. Intrinsic apoptosis is a common defense against cancer, while extrinsic apoptosis is often a response to viral infection. Both result in a very similar sequence of events leading to cell death in a more orderly fashion than the alternatives, with less potential for subsequent damage to other tissues and organs.

There are several ways in which cancer starts. Some people are born with mutations or variants of genes that increase the odds of developing cancer. For example, specific versions of the BRCA1 gene can leave women with a very high, inherited susceptibility to breast cancer. Also, some viruses can cause cancer. Human papillomavirus (HPV) infection increases the odds that a woman will develop cervical cancer by introducing viral genes that promote uncontrolled replication of the infected cells.

Further, we are constantly bombarded by environmental factors—toxins and radiation—that can damage our DNA, potentially leading to cancer. Little pieces of DNA in our genomes are being changed all the time. Errors are made during DNA replication, and molecular reactions can alter our DNA sequence. We have mechanisms for DNA repair, but if DNA damage happens at the wrong place and the wrong time, cancer can result. There are two ways that DNA damage can affect function of the protein encoded by a specific gene. Some mutations make a protein nonfunctional, rendering it unable to do its job, while other changes in the DNA sequence actually result in a mutant protein that is hyperactive.

Some cancers happen because the mechanisms that are supposed to stop them fail. We have genes called tumor suppressors that encode proteins that prevent cancer. These proteins make sure that cellular replication only happens when and where it is appropriate.

They are the brakes that keep the car from accelerating out of control down the mountain road. However, sometimes cancers are able to grow because inactivating mutations occur in tumor suppressors—as if some villain had cut the brake lines so that the car can't be stopped.

On the other hand are oncogenes, which normally promote cell replication. Mutated versions of oncogenes that display increased, uncontrollable activity can promote cancer formation. In this case, it's as if the accelerator in the car got stuck. Mutated oncogenes put the pedal to the metal, as it were. An example of an oncogene is a mutated growth-factor receptor that continuously signals to the cell that it is time to be fruitful and multiply, even in the absence of an actual growth factor.

It should be noted that tumor suppressors generally do a very good job. If an oncogene mutation arises, usually the cell can prevent cancer from forming. Similarly, if a tumor suppressor is inactivated, there still needs to be some sort of trigger, an oncogene mutation that will promote increased cellular replication. These observations led cancer geneticist Alfred Knudson in 1971 to suggest the "two-hit hypothesis," which basically says that both a tumor suppressor and an oncogene need to be mutated for cancer to actually emerge. You need to slam down on the accelerator *and* cut the brake lines to lose control of the car. However, as a last line of defense against these kinds of DNA damage, mechanisms have evolved by which the cell destroys itself before cancer can result.

There is a protein called p53 that sits at the crossroads of DNA damage, apoptosis, and cancer. The p53 protein is a tumor suppressor that stimulates a cell to enter apoptosis if extensive DNA damage is detected. However, this protective shield can fail. More than half of all human cancers are thought to involve mutations in the p53 tumor suppressor itself. That aside, let's take a look at how the normal protective mechanism works—how DNA damage leads to apoptosis.

When p53 recognizes a level of DNA damage that is dangerously high, it increases the expression and activation of a family of proteins known as Bcl2. Two members of the Bcl2 family, Bak and Bax, then stick to the outer membranes of mitochondria, which results in disruption of the mitochondrial permeability barrier. More simply put, holes form in the outer mitochondrial membrane so that substances that are normally kept inside of the mitochondria leak out. This is one of the main signals that activate apoptosis.

What actually happens during apoptosis?

There is a series of degradative digestive enzymes called caspases that lie in wait, unable to cause any damage until activated. When apoptosis starts, via any trigger or pathway, intrinsic or extrinsic, activating the caspases is the final common pathway that leads to the destruction of the cell from the inside out. Apoptosis is a bit like breaking camp on a scout trip. The actin microfilaments that hold the cell's shape, rather like tent poles, are degraded, so that the cell collapses upon itself. Also, the nuclear envelope disappears, and the DNA is degraded. In the end, there are only small remnants of the cell, referred to as apoptotic bodies, which are ultimately devoured by the ravenous macrophages.

The process of extrinsic apoptosis occurs in response to viral infection or other changes that are detected by surrounding cells. As noted, most of the final, destructive events are similarly orchestrated in intrinsic and extrinsic apoptosis. It's the triggers that are different. Extrinsic apoptosis is often imposed on infected cells by the immune system. There is a class of white blood cells called killer T cells. The "T" refers to the cells' origin in the thymus gland, so they are like the helper T cells that orchestrate immune responses, only a lot less friendly. Killer T cells rove around our bodies like assassins looking for double agents that need to be taken out. If a cell has been "turned" by a viral infection, if it has ceased to be a member of our team and now owes allegiance to the tiny viral invaders, the mission of the killer T cells is to terminate it with extreme prejudice.

Once infected with a virus, the cell becomes a little virus factory. This is why viruses invade our cells—it's how they reproduce. The infected cell uses the mechanisms of gene expression, including the secretory pathway, to generate new virus particles. Throughout the process of infection, viral proteins end up displayed on the surface of the cell. This is a huge red flag for the killer T cells.

All of our cells contain a pathway for shuttling little bits of proteins out onto the cell's surface like a white flag, telling the immune system, "We're on your side—don't shoot!" This pathway is based on what is known as the major histocompatibility complex (MHC). Scientists gave it the name "histocompatibility" because this is also the process that tells our immune system that foreign tissue is inside our bodies—it's what causes "rejection" of transplanted organs. (MHC was first described by George Davis Snell, who was awarded the 1980 Nobel Prize in Physiology or Medicine for his work, although Peter Gorer also played a significant role.)

Normally, the MHC on the surface of our cells is holding up white flags, bits of our own proteins, signifying to our immune system that the cell is a "friendly." However, in cells infected by a virus, the MHC displays viral proteins. The killer T cells see these red flags and act to remove the infected cells. (In the case of transplantation, the donor's MHC itself is seen as foreign and this really sets off the immune system.)

In a fiendishly ingenious bit of evolutionary trickery, some viruses have actually evolved mechanisms to prevent the MCH pathway from functioning, thus enabling themselves to hide in plain sight from immunological surveillance. HIV expresses a protein called Nef that functions to manipulate the vesicle trafficking system of the infected cells so that the presence of MHC on the cell surface is reduced.

There are two different ways in which killer T cells act to destroy cells. Killer T cells contain lytic granules, little hand grenades that can be tossed at infected cells nearby to destroy them. Killer T cells can also fight viruses by activating the extrinsic apoptosis pathway. Every cell has a self-destruct button on its plasma membrane, a receptor known as Fas. When Fas is activated, the cell will enter apoptosis. Like good commandos, killer T cells know where these self-destruct buttons are, and have orders to push them if they find bits of viral proteins displayed in the cell's MHC. This results in activation of caspases and destruction of the infected cell.

So the cell is always in an evolutionary arms race, fighting to develop mechanisms that will protect us from infection and cancer, just as the pathogens and cancer cells are being selected for the ability to survive and grow.

Immune cells themselves can be triggered to undergo apoptosis, which can reduce the activity of our immune system, not always as bad an idea as it sounds. This can be beneficial in preventing autoimmune diseases and even transplant rejection. However, in other cases the dulling of an immune response can be harmful. Cancer cells have disorganized genomes and improper regulation of protein expression. That means that proteins not normally expressed by mature cells, or those that are usually expressed in only very low amounts, can be found at high levels in cancer cells. This can mark cancer cells as different from other normal cells, helping our immune cells target them for destruction. However, cancer cells can also express signaling proteins that promote apoptosis in immune cells, turning the tables against the good guys. Therapies are currently being developed that will inhibit immune-cell apoptosis, in order to strengthen an immunological response to cancer. These types of approaches carry significant risk, as negative regulation of immune-cell function is very important to prevent inappropriate immunological responses. Nonetheless, inhibition of T-cell apoptosis has been shown to extend the lifespan of people with certain types of cancers.

CHAPTER 19

The Mystery of HIV

◉

I N PREVIOUS CHAPTERS WE'VE TOUCHED ON THE MATTER OF HOW viruses need to invade host cells in order to reproduce. Exactly how viruses get inside cells is a fascinating field of study, one that bears on successful and potential treatments for viral diseases. The most infamous virus of our time, human immunodeficiency virus (HIV), is a particularly enigmatic case with regard to how it invades cells. While medical science has made great progress in keeping HIV patients alive, we still don't really know how the virus gets inside cells, even though that knowledge could be useful in developing ways to prevent or cure the disease HIV causes—acquired immune deficiency syndrome, or AIDS.

In chapter 5 we mentioned that some viruses' genomes contain DNA and others RNA. There are additional ways in which viruses are classified. "Enveloped" viruses are bounded by a lipid membrane in the same way as our cells are. The viral envelope surrounds and protects the inner protein shell, called a capsid, which contains the genetic material. There are also many non-enveloped viruses, essentially a naked capsid floating around, hoping to get brought into cells. Viruses in this class include the human papillomavirus (HPV), which can cause cervical cancer, as well as the rotavirus responsible for upset stomachs the world over.

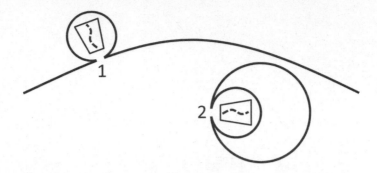

In order to infect a cell, an HIV particle consisting of the
viral capsid and RNA genome needs to get inside and get
rid of the particle's "viral envelope" by fusing it with a
cellular lipid membrane. There are two principal models
for how this happens: (1) HIV fuses directly with the
plasma membrane of a target cell; (2) the virus enters the
cell via endocytosis and then fuses with the membrane of
an endosome formed during the process of endocytosis.

The membrane surrounding enveloped viruses contains surface
proteins that interact with the outside world. In fact, the viral envelope is actually just a bit of the plasma membrane from the cell off
which the virus particle budded. Many of the viruses that infect
humans are enveloped, including HIV, Ebola, SARS, herpes, influenza, and hepatitis. The envelope plays a key role in fusing the virus
to our cells, introducing the viral proteins and genome to the intracellular space.

Lipids are generally relatively simple molecules that have very
special properties that make them ideal for forming the protective
membranes around cells, subcellular compartments, and enveloped
viruses, basically anywhere you want to separate two fluid-filled
spaces. Lipids (which are essentially fat or oil molecules) have two
parts: a "head group" and a "hydrocarbon tail." The head group contains elements like oxygen and nitrogen that tend to bind to water
molecules, a tendency referred to as hydrophilicity. The hydrocarbon
tail is made up solely of hydrogen and carbon. Hydrocarbons resist

interaction with water and thus are referred to as hydrophobic. We often hear the word hydrocarbon in the news to describe fossil fuels, coal, natural gas, and oil, and of course, we all know the old adage "oil and water don't mix." The same is true of the hydrophobic tail of a lipid molecule—it really hates being in contact with water. So when lipid molecules group together to form various kinds of structures, they do so in ways that keep those hydrophobic tails dry, as it were. These self-assemblies include starlike shapes called micelles, in which all the tails point inward to avoid exposure to water, while the heads point outward, maximizing contact with water. Another shape that can form under the right conditions is referred to as a "bilayer." As you might guess, a bilayer consists of two layers or sheets of lipids plastered together with the head sides out and tail sides inside, facing each other. In this way, a bilayer functions as a barrier.

All cells are bound by a lipid bilayer, the plasma membrane we've discussed through this book, a bit like a water balloon or the little plastic bag in which pet stores put goldfish for you to take home. In fact, the dimensions, the ratio of the bilayer thickness to the size of the cell, are about the same as in a water balloon. Some cells, especially plant cells, have an additional, harder, protective "cell wall" made of carbohydrates.

A single HIV virus particle is 120 nm across and comes wrapped in a viral envelope, a piece of bilayer membrane about 10 nm thick. So from the virus's point of view, the challenge isn't merely getting inside the cell. It also needs to shed its envelope—check its coat at the door—for the infection to be successful, or "productive" as the virologists say.

As with other retroviruses, the HIV needs to use reverse transcriptase to convert the viral RNA to DNA and then integrate its genetic material into the genome of the host cell in order to complete its life cycle. This is only possible if the viral envelope is removed. The way this generally happens is by merging, or fusing, the viral envelope with a lipid bilayer of the host cell, effectively removing the envelope from around the protein-based capsid and genome within.

The commonly accepted mechanism for how this happens is "direct fusion," in which the viral envelope merges directly with the plasma membrane of the target cell, permitting the infectious core of the viral particle access to the interior of the host cell.

HIV contains a protein on its surface called GP120 that binds to a protein on the surface of the target cell called CD4. CD4 is an important factor for cells of the immune system, in particular, helper T cells, which act as important regulators of our immune responses. It is because HIV specifically targets our immune cells that it is such a dangerous pathogen. Over time, as the "viral load" of an infected person goes up, helper T cell numbers go down, and the person's immune system stops working. In fact, it is not the HIV virus, or AIDS, that kills people per se. Rather, by lowering helper T cell levels, it robs the body of its ability to respond well to other infections. So it is usually something like pneumonia that kills a person with AIDS, as the victim's immune system can no longer fight off common kinds of infections.

In the direct fusion model, the GP120 on the HIV particle binds to the CD4 on the surface of a helper T cell. At this point, in some way not well understood, the viral envelope fuses with the plasma membrane of the host cell. This is something that has reportedly been observed by electron microscopy, but not often or convincingly. Thus, some scientists believe HIV probably uses a different mechanism to infect cells. An alternative theory suggests that, rather than direct fusion at the outer surface of the target cell, the virus is taken into the cell intact and *then* fuses with the membrane surface of an organelle inside the cell known as an endosome.

This model involves endocytosis, and understanding it requires a brief diversion to explain what that is. Very small things, like ions (charged atoms or molecules), enter and exit cells via small pores in the plasma membrane called channels or transporters. These act like little pipes, in which opening and closing the valves is tightly regulated. Larger substances that need to get in or out of the cell have to use far more complicated and indirect means of transport.

Endocytosis is the process by which this stuff gets in from the outside. Cells internalize fluid and dissolved substances by pinching in small patches of the plasma membrane and turning them into little "beggar's purses," vesicles that contain the cargo and a bit of the fluid in which the cargo was floating outside the cell. The usual kinds of cargo are things like nutrients or growth factors, which bind to receptors on the cell surface and switch on different activities inside the cell, such as expression of particular genes. However, endocytosis is also a way in which pathogenic viruses can hitchhike their way into target cells.

That is the basis for this alternative model of HIV entry—the endocytosis of viral particles following binding of GP120 to CD4. Internalization of cell surface receptors such as CD4 routinely occurs by endocytosis, and we know that other viruses, such as the influenza virus, enter cells this way and fuse with membranes inside the cell. Following endocytosis, the cargo brought in percolates through a series of compartments in the cells—endosomes. These are interconnected network way stations throughout the cell. A vesicle containing cargo will fuse with an endosome, and then new vesicles will bud from that endosome and carry the cargo through the cell. As cargo moves, carried through this network from early endosomes to late endosomes, the fluid inside the endosomes becomes more and more acidic. The endosomal pathway ends up in the lysosome, which is a highly acidic compartment for degrading cellular waste. (In the previous chapter, we likened it to a sewage-treatment plant.) So if an infection is to be productive, the virus has to escape from the endosomal pathway before it reaches the lysosome, or it will be destroyed.

As proteins move from neutral to acidic environments, their structures change. The influenza virus makes ingenious and insidious use of this phenomenon. Once an influenza virus particle has been delivered to acidic endosomes, a protein in the viral envelope called hemagglutinin, or HA, changes its shape in a fascinating and potentially deadly way. Its new shape enables the HA protein to insert itself into the endosomal membrane so that the viral envelope fuses with the inside of the endosome, and the newly unveiled viral core pops out of the endosome into the intracellular space. Just getting the virus inside the cell is not the culmination of the viral life cycle. The virus must get the host cell to make more viral particles. This is the difference between simple, or not so simple, viral entry and true infection. In order for an influenza infection to be truly productive, the genetic information encased in the viral capsid must make its way into the nucleus of the host cell. Like a molecular Swiss Army Knife, the HA protein changes its shape and function in response to the acidic environment of the endosome, creating the route to infection. Thus, following escape from the endosome, the exposed virus will then make its way to the nucleus.

It follows from this that drugs that remove the acid from the endosomes might inhibit influenza infections. In fact, such drugs are currently being tested as potential treatments for influenza. Likewise, we've known for a long time that HIV is also found within endosomes. However, these same influenza drugs have no effect on HIV. So it was assumed that the presence of HIV in endosomes was an epiphenomenon, a scientific term for a red herring.

In short, it was thought that the endosome was a dead end for HIV particles. But more recently, the dogma of direct fusion has been challenged by a series of experiments that have directly tested whether HIV entry by endocytosis can result in productive infection. This work is still in its infancy, though, and it is not yet generally accepted that in the human body HIV productively infects helper T cells via endocytosis rather than via direct fusion at the plasma membrane. This remains a very exciting time for this area of research, which is clearly of key importance to our understanding of the greatest viral pandemic of our time. However, at this writing, we really have no idea how HIV gets into our cells to infect them.

PART FIVE

CELLS, ORGANS,
AND ORGAN SYSTEMS

CHAPTER 20

What Are Organs and Why Do We Have Them?

◉

C ELLS ARE OFTEN STUDIED IN ISOLATION, AND THAT IS PRETTY much how we've discussed them so far. In the body, however, cells are generally the basic components of tissues, organs, and organ systems. Cells must interact with each other in order for these collective functions to emerge. In many cases, cells communicate with one another via proteins at the plasma membrane, as you've already learned. Integral membrane proteins at the cell surface serve as receptors for molecules that are released from one cell to signal to others, as well as for other membrane proteins in cells to make direct contact. You've heard of endocrine signaling, in which specialized cells and organs release hormones that move through the bloodstream, often to distant targets. The pancreas, for example, secretes the hormone insulin into the blood, which regulates the body's ability to process and store sugar. On the other hand, "paracrine" signaling occurs when one cell secretes a factor that binds to a receptor on a nearby or adjacent cell.

In addition to binding to small secreted factors, membrane proteins in adjacent cells stick to each other. Moreover, signaling is not the only result when adjacent cells interact via cognate (i.e., corresponding) membrane proteins. Junctions that form across the extracellular

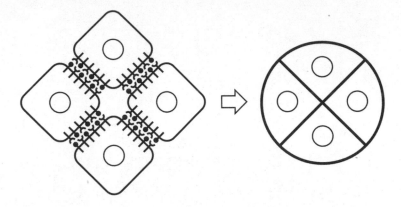

Binding of cognate cell-adhesion receptors between adjacent cells leads to the formation of cell-to-cell junctions necessary for organ development.

space can be strong and long-lived. This, in fact, is how cells adhere to one another to form organs and tissues. Interestingly, signaling and cell adhesion are often interdependent, and the junctional proteins are often also signaling receptors. This is how the cell *knows* when it is bound to others in collective clusters. The signaling status of the cell can alter its adhesive properties, and when cells meet and stick to each other through adhesive cell-surface proteins, this signals a slowing down of replication. So-called "contact inhibition" was discovered by Michael Abercrombie in the mid-1960s and very soon after, it was noticed that while normal cells stop actively replicating once they make contact with their neighbors, cancer cells do not. When cells are overstimulated so that growth-factor signaling progresses unhindered, cancer results. One of the main hallmarks of cancer is the loss of normal interactions between cells. Cells lose their contacts with their neighbors and replicate in an uncontrolled manner beyond the confines of the normal organ system from which they originated.

Cells in organs are often surrounded by connective tissue. Cells called fibroblasts secrete extracellular matrix (ECM) molecules. The fibroblast acts like a spider at the center of a three-dimensional web of ECM in which other collections of cells can become stuck, like flies in the web. In this way, different substructures and layers of cells can form together into tissues, organs, and organ systems. (Also, ECM fibers are a route by which metastatic cancer cells can travel to get out of the organ in which they developed and create new tumors elsewhere.)

There are many different types of organs, and although some are self-contained and perform specific defined roles at particular times and places, others are distributed networks harder to imagine as a single functional unit, such as the circulatory system. In the former category are, for example, the heart, lungs, liver, and kidneys. Each of these is essentially a grouping of particular cell types that work together to accomplish specific tasks. Many of these solid organs do something to our blood. The heart pumps the blood around the body. The lungs add oxygen and remove carbon dioxide. The liver uses enzymes to remove toxins and to process and store certain substances that come from our food. The kidneys filter our blood, removing waste products and balancing the amounts of salt and water in our bodies. In each case, these organs need their own blood supply, as well as a conduit through which blood can flow in order to be processed or acted upon.

Our circulatory system consists of vessels through which our blood flows. Blood leaves the heart rich in oxygen and is delivered to organs and tissues via arteries. After oxygen is delivered, blood returns to the lungs through veins. Our arteries and veins vary considerably in size, depending on the volume of blood required to flow at a particular location. The aorta, which is connected directly to the heart, is the widest conduit for blood in the human body—it can be over an inch in diameter. At the outer reaches of the circulatory system are the tiny capillaries that interlace through our organs and tissues and can be as little as 1/10,000 of an inch in diameter. Capillaries are often so narrow that only the smallest red blood cells and platelets can enter and flow through them. Capillaries are like small single-lane dirt roads in the country, while the aorta is like a super-highway traveling through the heart of a city. (In certain patholog-ical states, the aorta can grow even larger, forming an aneurysm, which can tear with disastrous results.) At any rate, blood vessels, skin, and the nervous system are examples of similar cell types that can be defined as organs, even though they are distributed through-out the entire body. Even the blood is a sort of organ, actually two distinct ones.

The blood serves many roles, including the delivery of oxygen and nutrients and the transport of carbon dioxide and waste products. You will remember from previous chapters that energy is produced by the mitochondria in the form of ATP. This, like any process of burning fuel, consumes oxygen and produces carbon dioxide. Similarly, waste products rich in nitrogen, such as urea and uric acid, are produced from the breakdown of proteins and nucleic acids, respectively.

The circulatory system also serves as the conduit for our immune system. It's the way our bodies' "first responders" travel to any part of the body that has been damaged or is under attack from foreign invaders such as viruses or bacteria. In fact, we have two parallel cir-culatory systems. One is the arteries and veins that carry our blood. The other is the lymphatic system, which transports white blood cells. The lymph acts as the service road to the cardiovascular system's highway. Lymph is formed from fluid that drains out of our organs that is collected and moved through the lymphatic system, ultimately joining our bloodstream via the subclavian vein near the collarbone.

Our immune system includes many kinds of cells that work together to be on the lookout for pathogens and mount a response if any are detected. Many immune cells are found at common sites of pathogen entry, such as our lungs and nasal cavities. However, because infection can spring up anywhere, white blood cells reside in nearly all organs and tissues. In addition to these agents in the field, white blood cells circulate between our tissues and various immune-system organs. Cells of the immune system are generated either in the bone marrow, along with red blood cells, or in the thymus, the site where T cells originate—both the helper and killer types. The spleen serves as a central way station where red and white blood cells develop, interact, and are stored, sort of like the CIA headquarters in Langley, Virginia. Our lymph nodes are localized self-contained spaces where different immune cells can communicate and become activated, like CIA field offices in various parts of the world. When an infection occurs, the local lymph nodes become inflamed as sentinel cells report back their findings, and reactive immune cells replicate and activate.

A critical component of the immune system is the ability of the cells that first sensed and responded to an infection to communicate their findings to other cells that orchestrate body-wide reactions. Also, certain immune cells will be stored following an infection, preserving a memory of that particular response, so that if the same pathogen strikes again, the immune system can recognize and eliminate it more rapidly in the future. This is sort of like retired field agents that can be used as analysts to assess potential threats and assist in developing a more efficient strategic reaction the next time the same enemy attacks.

On the other hand, what we most readily think of as organs are more localized in structure and more specific in function, sort of like Pittsburgh when it made steel or Detroit when it was the automotive capital of the world. The cardiovascular system functions like highways, rivers, and railroads, bringing raw materials in and carrying the products out. Like cities, organs can be slow and even resistant to adjusting their standard structures and functions in periods of change, stress, and disease. Most solid organs do what they are designed to do with an inherent ability to modulate their functional status or output within specific ranges. However, like twenty-first-century Detroit, they suffer from a limited capacity to respond to changes imposed from outside forces. If they are subjected to conditions beyond their capabilities, they will rapidly cease to function effectively. Thus, solid organs can be described as more "binary" than "digital." They are not governed as part of an interactive network. Rather, they are subject to rigorous regulations of limited flexibility.

Physiology is the study of how the body and its systems respond to changing conditions. An organ such as the kidney lies at the center of a complicated web of physiological regulations. However, this can ultimately be understood as a series of linear, causal relationships that can be identified and studied. One of the main challenges that any organ can face centers around the limits being imposed on its blood supply. The heart, for example, is primarily composed of muscle, which requires a constant supply of oxygen and energy. However, poor circulatory health, such as the presence of blockages made of fats that have not been properly processed or stored, can result in reduced flow through the coronary arteries, whose job it is to keep the heart tissue healthy and functioning properly. When the flow of blood that delivers oxygen to the heart tissue is restricted by such a blockage, the cells of the cardiac muscle begin to die, resulting in a heart attack and possibly causing death. Therapeutic interventions can be implemented to increase blood flow to the heart muscle. There are drugs that reduce the deposition of fat along the arterial wall to prevent blockages. In the procedure called angioplasty, which has saved countless lives, doctors insert a long, thin catheter that has a small balloon at the end into clogged coronary arteries. The balloon is inflated to push open the constricted part of the artery. Balloon angioplasty can be risky, as the procedure is often traumatic to the coronary arteries and can actually engender blood clots formed of small blood cells called platelets. In the mid-1990s, Barry Coller, now the physician-in-chief at the Rockefeller University in New York, developed Abciximab, which is sold under the commercial name Reopro. This is an agent that prevents platelets from sticking to each other, and it has significantly reduced the rates of complications following catheterization procedures such as angioplasty.

In cases where blockages cannot be cleared with drugs or angio-plasty, surgery can be performed. One type of heart surgery involves the removal of blood vessels from one part of the body, often the leg, which are then sewn on to either side of the blockage to allow blood flow to bypass the site of the obstruction. Coronary-artery bypass surgery has been performed millions of times since it was developed in the 1960s. However, the ability to sew blood vessels together was developed by Alexis Carrel half a century earlier. Carrel was a true surgical innovator who standardized a series of astounding feats of surgical skill, some of which he learned from seamstresses, that earned him the Nobel Prize for Physiology or Medicine in 1912 and are still in use today.

If blood flow is restricted, cells suffocate from lack of oxygen. Flow through the tissue ceases, causing a buildup of waste products, and cell death rapidly results. This is why heart attacks and strokes, blockages to the blood supply of the brain, can kill so quickly. How-ever, while restoring blood flow to damaged organs rapidly can mean the difference between life and death, as we will learn later there are instances when strangling off the blood supply to cells is a therapeu-tic imperative, in particular in attacking a tumor. However, first we must take a detour into a critical organ that regulates blood pressure and salt and water balance. Feel free to put down this book for a momentary "comfort break" if you need.

CHAPTER 21

The Kidney:
Cells in Concert

◉

M ANY ORGANS ARE HIGHLY ORGANIZED, SELF-CONTAINED structures formed from integrated populations of specific cell types working in concert to orchestrate complex chemical changes. Just as a philharmonic orchestra is arranged with the various kinds of strings, woodwinds, brass, and percussion grouped together, cells of distinct types within an organ are grouped by function for optimal performance, and these different "sections" work interdependently as an ensemble.

These different groups of cells are given their cues to begin or stop playing, to play faster or slower, louder or softer, by hormones that are produced and released by different glands throughout the body and carried through the bloodstream until they reach a particular organ. Like the orchestra's conductor, these hormones regulate what the organ produces and the strength and duration of that output. Adrenaline, for example, stimulates blood flow and increased metabolism, like a conductor using a baton to increase the tempo from andante to allegro as the sheet music requires.

Each type of organ has its own unique combination of component parts. As with different kinds of musical ensembles—a klezmer band or a jazz combo or a zydeco or bluegrass group—the specific

173

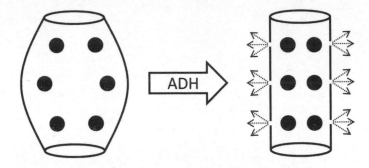

The kidneys filter waste products out of the blood
and regulate the content of salt, potassium, and other
substances (black dots). Antidiuretic hormone (ADH) helps
regulate fluid levels by promoting insertion of aquaporin
water channels into the plasma membrane of principal
cells in the collecting duct. This duct allows water to exit
the kidney back into the blood rather than flow into the
bladder, decreasing urine volume and increasing urine
concentration.

composition of instruments and parts determines the output. The
spleen, liver, and kidney modify the composition of blood in different
ways, even though all contain some similar types of cells.

The kidney is a great example of numerous semiautonomous cell
groupings working together to produce an ever-changing output, in
this case, not music but urine.

Basically the job of the kidney is to filter our blood, returning the
blood cells to the general circulation while holding on to the filtered
fluid that is left over and regulating its composition and concentration, preparing it for excretion in the form of liquid urine. In addition
to removing waste products, the kidney also regulates salt and water
balance. If you have ever taken a diuretic, or have drunk a lot of coffee or beer, you've surely noticed an increase in the volume of urine
you produce. And as the volume goes up, the concentration of waste
molecules dissolved in it generally goes down.

Urine is a solution, made up of a solvent, essentially water, and solutes, things like salts and waste products. You may have noticed that when you are dehydrated, when you haven't been drinking enough water, you tend to excrete a low volume of brightly colored, highly concentrated urine. Likewise, in the previous instance, where you had been drinking a lot of coffee and producing a lot of urine, the color was pale or even clear. This illustrates the plasticity in the kidney's ability to concentrate or dilute urine as needed. If you increase the amount of solvent in a solution, its volume will go up and the concentration will go down. Furthermore, since blood is the source of the fluid that becomes urine, the more urine we excrete, the lower our blood pressure; the volume of blood and of urine exist in balance. This is why diuretics, substances that make us excrete large volumes of dilute urine, can be used to regulate hypertension, i.e., high blood pressure.

The kidney is made up of tiny tubes called nephrons. On top of each is a tiny molecular sieve called a glomerulus, creating a filtered solution devoid of blood cells that enters the nephron. This is why doctors who treat kidney disease are called nephrologists. There are about a million nephrons in the human kidney. Each is about an inch or two long, but, obviously, they are incredibly small in diameter. In fact, each nephron is made of a single layer of cells wrapped into a tube. The cross-section of a nephron may be as narrow as 20 microns across, as it can be composed of as few as three or four cells wrapped in a tube around a fluid-filled space. Thus, nephrons are somewhat like the small capillaries that blood flows through in our tissues. However, the fluid that moves through nephrons is quite different from blood, and the nephron itself plays a very active role in controlling the composition of the fluid that ultimately becomes urine. Unlike a blood vessel, a nephron is not a passive conduit but actively regulates the liquid by adding and removing substances as it travels through on the way to the bladder.

Different segments of the nephron have different functions. Some areas regulate the specific ions and waste products that are present in the fluid, while others control the general concentration and volume by modifying the amount of water the nephron contains. Ions such as sodium, calcium, and potassium can be transported into and out of the fluid inside nephrons. In fact, because too much potassium in our blood can be dangerous, people with kidney disease are generally advised to avoid potassium-rich foods like bananas. The composition of the fluid in the nephron, and ultimately in urine, is controlled by proteins present in the plasma membranes of the nephron's cells.

Specific transporters and channels in the surface of these cells regulate the amounts of particular substances in our urine. Different cells along the nephron express different types of membrane proteins, and these function together in a complex chain of events to ultimately generate the urine composition required, depending upon diet, water intake, and other metabolic and environmental factors.

There are also cells that specifically function to regulate the acidity of urine. These are called intercalated cells, and there are two kinds. The α-intercalated cells acidify the urine if it is too basic, not acidic enough, while β-intercalated cells make the urine less acidic, more basic. Type α-intercalated cells contain transporter proteins that add protons to the fluid inside the nephron to make it more acidic. Type β-intercalated cells pump bicarbonate into the fluid. Bicarbonate neutralizes acid in the same way that chewing Tums, which is made of calcium carbonate, helps when you have an acid stomach. Remarkably, α cells can become β cells, and vice versa, when necessary due to protracted acidity or basicity. So the kidney is a very plastic organ. It can easily adapt to changes in the composition of blood and, ultimately, urine.

An amazing example of this plasticity exists near the end of the nephron, where it approaches the bladder. This area regulates the amount of water in the fluid in the nephron, and thus the volume and concentration of our urine. When we are dehydrated, our kidney responds by reducing the volume of urine excreted and thus increasing the concentration. This response is triggered when the pituitary gland releases antidiuretic hormone, also known as vasopressin, at times when the body needs to conserve water—for example, if our blood pressure drops. Upon reaching the kidneys, antidiuretic hormone acts upon what are called principal cells, which are found in the collecting duct. Specifically, it increases the presence of water channels called aquaporins in the surface of principal cells facing the inside of the nephron. This enables more water to flow out of the nephrons before it can reach the bladder, thus decreasing urine volume and increasing concentration. In fact, some substances that act as diuretics function by inhibiting aquaporins.

Aquaporins were discovered by Peter Agre, who, at this writing, still works at Johns Hopkins. Aquaporins are not only found in the kidney but exist in nearly all cells to regulate cellular fluid balance. Agre discovered aquaporins by accident when he was studying membrane proteins in red blood cells that are responsible for whether people are Rh-factor positive or negative. Instead, he identified a fundamental membrane protein that is absolutely required for life in animals, plants, and even bacteria. For this pioneering work, Agre was awarded the Nobel Prize in Chemistry in 2003, and the fact that he made this discovery while looking for something else doesn't diminish its importance or Agre's achievement.

When you're drinking enough water, the absence of antidiuretic hormone reduces the presence of aquaporins in the plasma membranes of the principal cells so that more water flows into the bladder as urine. In general, the regulation of kidney function is conducted via linear systems subject to straightforward rules of cause and effect. When changes in urine composition are required, the function of the kidney changes. While elegant and fascinating, these systems of detection and response are relatively direct. This is why many of the functions of the kidney can be replaced by dialysis, in which a machine filters blood to remove toxins and regulate salt and fluid balance.

We have an amazing understanding of how most organs in the body are regulated—the kidney, the heart, the lungs, the liver, and others. In each case, we understand how the needs for changes in function are sensed, what hormones are involved, and how the organs respond. The instant we sense danger—when we see a lion or (more likely for most modern humans) a bus hurtling toward us—our adrenal gland instantly releases the hormone adrenaline, which increases our heart rate, blood pressure, and respiration in support of the "fight or flight" response. (Cautionary note: in the case of either the lion or the bus, choose flight.)

This is only one simple example, but the functions of many organs can be described in terms of systematic cause and effect and the integrated functioning of various cell types that make up the organ. However, there is one organ that has long eluded our understanding. Ironically, it is the organ in which the ability to understand things resides—its network-like organization is constantly changing, and it is so complicated in structure and function that its function cannot be reduced to a model. This organ is, of course, the brain, and although we are finally beginning to understand how the central nervous system works, we have taken only our first baby steps toward a truly integrated description of how things like learning, memory, and personality are created.

CHAPTER 22

The Brain: How It Works, How We Learn, and Why It's Hard to Study

B RAIN CELLS AND THE OTHER CELLS OF THE NERVOUS SYSTEM are called neurons. Neurons come in a wide variety of shapes and sizes. Some are as small as a few microns across, while others can be as long as a meter, such as those that run from the base of the spine all the way to our feet. Some are shaped like little stars, and are actually called stellate neurons, while many are very long and thin. There are as many as 100 billion neurons in the brain, but neurons also make up the network of nerves that runs throughout the entire body. Most neurons are organized with three main parts: the axon; the cell body, or soma; and the dendrite. The axon is generally a long shaft extending off of the neuron. It is the front end of the cell, and when a neuron moves from one place to another—for example, during development—the axon drives the neuron forward. In fact, "axon guidance" is a major area of study, as it underlies development of the nervous system and could be harnessed to promote repair in cases of damage or disease. The axon regulates the ability of a neuron to move, and contains much of the machinery that controls the process of cell migration, including the ability to sense direction correctly.

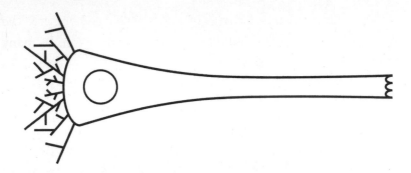

A typical neuron where electrochemical signals travel from the branched dendritic network on one side of the nucleus through the long, thin axon that terminates in the growth cone on the other end.

In the late nineteenth century, the neuroanatomist Santiago Ramón y Cajal discovered that at the very tips of growing axons are small protrusions referred to as growth cones, which contain the receptors and adhesion proteins required for axon guidance. The receptors detect chemical signals that mark the ultimate destination of the motile neuron, while the adhesion molecules link the moving cell to the surrounding environment. Cajal originally wanted to be an artist, and his talents in the area are evident in the amazing drawings that he made of neurons. Instead, he followed in the footsteps of his father, who was an anatomist. In addition to first describing the axonal growth cone, Cajal proposed that specific chemical signatures in the brain drove neuronal migration in a process similar to that which drives immune cells toward sites of infection (chemotaxis). Cajal was awarded the Nobel Prize in Physiology or Medicine in 1906 for his work. In fact, he shared the prize with Camillo Golgi, after whom the Golgi apparatus in the secretory pathway is named.

The cell body of a neuron lies just behind the shaft of the axon and contains the nucleus and many other important cellular organelles. A large array of parallel microtubules extends out from the cell body that carry organelles and vesicles from the cell body through the axon.

At the rear of the neuron, behind the axon, is the dendrite. The dendrite is generally not a single protrusion like the axon. In many neurons, a branched dendritic network spreads out in all directions behind the cell body. In some neurons, this structure is so intricate that it calls to mind the organization of a leafless tree; it is thus referred to as the dendritic arbor.

The structure and function of a neuron are intricately linked, so before we consider what these dendritic extensions actually do, we need to discuss what neurons themselves do.

The simplest answer is that they transmit information. However, in addition to merely transmitting signals, they work together to store and process information. These functions depend on the connections between neurons, and many neurons maintain thousands of connections with others around them. These connections, called synapses, are how signals are passed from one neuron to another. Each neuron can have thousands of synapses, and the human brain can have hundreds of trillions of synapses. The architecture of this network, how and where different neurons are connected together, is the basis for the transmission, processing, and storage of information—and so, for how we think, feel, remember, and react, among other things.

The basis of synaptic transmission is that the axon of one cell connects to the dendrite of another. This involves physically stitching together the axon of cell 1 with the dendrite of cell 2. These connections involve cognate cell-adhesion proteins on the two opposed neuronal membranes. These connect the two neurons together like molecular Velcro.

In most synapses, these connections are in a circular pattern, so that a very small space exists between the two neurons, sort of like two paper plates glued together with the curved spaces facing each other. Into this space, the presynaptic cell (cell 1) secretes molecules called neurotransmitters that bind to receptors in the membrane of the postsynaptic cell (cell 2). This starts a chain reaction called an action potential, which ultimately results in the release of neurotransmitters from the presynaptic terminal at the axon at the other end of cell 2, inducing another action potential in the next neuron in the chain (cell 3).

What happens when these neurotransmitters bind to their receptors in the postsynaptic terminals?

The inside of the neuron is generally negatively charged relative to the extracellular space. When neurotransmitter receptors on the postsynaptic terminal are activated, positively charged ions rush in through ion channels, temporarily making that part of the neuron locally lose its extreme negative charge. This then causes more positive charges to enter the cell through other ion channels in adjacent regions further down the neuron in a positive feedback loop that ultimately moves like a wave along the entire neuron. The first group of ion channels opens in response to neurotransmitter binding, while the second is actually regulated by the electrical conditions. This mode of regulation is what permits the positive feedback loop—it's how the wave of the action potential propagates along the entire neuron. The end result of the action potential is that the loss of the negative charge at the very tip of the axon causes secretion of neurotransmitters from the presynaptic terminal to activate an action potential in the next neuron in the chain.

The understanding of the action potential was experimentally and mathematically determined by Alan Hodgkin and Andrew Huxley, working primarily in Falmouth, Massachusetts, in the 1930s. This location was critical to their success, as they conducted their experiments on the "giant axon" found in squid brought in by local fishing boats. These giant axons can be up to half a millimeter in diameter, and in order for their experiments to work, the researchers needed to dissect them from freshly caught squid. The large size of these neurons permitted Hodgkin and Huxley to manually record action potentials with electrodes inserted into the giant axon. Hodgkin and Huxley published their recordings of the first action potential in 1939 but continued to work together for many years. In

1963, they were awarded the Nobel Prize in Physiology or Medicine for their insights into the electrical signaling of neurons.

A signal in the nervous system can originate from a large number of things. Many signals originate through our five senses. We have many different types of sensory neurons that detect changes in our environment and transmit these signals to the brain. These sensory neurons are stimulated and regulated via a wide variety of mechanisms, some direct and others indirect. Some sensory neurons contain the receptors used to detect environmental changes, while others communicate directly with sensory cells that contain the receptors, but are not themselves neurons. Sensory neurons in our eyes detect light directly. Those in our nose have receptors for odors. However, taste buds full of receptors in the tongue bind flavorful molecules in food and transmit signals to adjacent sensory neurons. More specifically, taste-receptor cells connect to neurons one layer down in the tongue that transmit signals that tell the brain what we are tasting. In the nose, however, dendrites of sensory neurons extend directly into the nasal cavity that contain the receptors that bind to odorant molecules present in the air we inhale. The activation of odorant receptors stimulates the sensory neurons to undergo an action potential that transmits the signal to the brain, which will recognize the signals as the scent of roses, the aroma of fresh-baked chocolate chip cookies, decaying garbage, and so on. Thus, the mechanism of detection and transmission in our sensory system is based upon many variations on similar themes involving receptors that detect signals and neurons that transmit them.

Although action potentials move like waves along the entire length of a neuron, the neurons are not actually completely covered by ion channels. The electrical pulse of an action potential actually moves along two different types of regions in our neurons. There are regions covered by a substance called myelin. This is literally electrical insulation, like the coating on a wire, and it increases the speed with which an action potential can move along a neuron. The myelin sheath is formed by Schwann cells, so called after their discoverer, Theodor Schwann. Schwann cells wrap themselves around the axons of a neuron, and this electrical insulation permits action potentials to rapidly move along the neuron without requiring the presence of ion channels along the entire length of the axon. Rather, the ion channels are concentrated in nodes between segments of the myelin sheath, like areas where the insulation has been stripped off of a wire. Thus, the action potential jumps from one node to the next extremely quickly, eliminating the need for ion channels along the entire length of the axon.

The presence of myelin is extremely important to proper neuronal function, and demyelination can be extremely dangerous, with catastrophic consequences for the nervous system. Multiple sclerosis is a debilitating neurological disease that is associated with demyelination. The causes of multiple sclerosis and other examples of demyelination can be varied and include genetic, environmental, and autoimmune components.

Oligodendrocytes are another type of cell that form myelin sheaths around neurons. While Schwann cells can only myelinate a single neuron, oligodendrocytes can contribute to the myelination of numerous nearby neurons. Furthermore, the function of oligodendrocytes can change over time, depending on specific conditions and stimuli. Recent evidence suggests that some types of learning—for example, a new complicated motor task—can depend on alterations in how particular neurons are myelinated, as this is one way in which particular pathways are reinforced. So learning how to ride a bike could involve oligodendrocytes actively remodeling the myelination of specific axons. This is why you might be very shaky at first, but eventually get the hang of it, and, as the saying goes, never forget how to ride a bike once you've learned.

Of course, myelination is not the only source of learning in the nervous system. The regulation of which neurons form synapses with each other and how strong those connections are underlies neurological development and learning in many ways. The processes governing the formations, rearrangements, and strength of particular synapses, called synaptic plasticity, are currently a very active area of study. However, the fundamental principles in this field were actually laid down over fifty years ago by psychologist Donald O. Hebb. Hebb's central hypothesis was very nicely paraphrased by neurobiologist Carla Shatz, who stated, "Neurons that fire together wire together."[1] What this means is that it is the function of synapses between cells that is the basis for learning and memory. Or, to reverse this theory, when we learn new things, the connections that are involved will be reinforced over time. When we take in new information or when the brain processes sensory stimuli, events, and actions that are important or repetitive, our synapses change so that those pathways and networks function more efficiently and are locked in for future use. This is one reason why you can perform the same task more quickly and deftly each time you do it. And it's why repeated use of flash cards helps us memorize vocabulary words when learning a new language.

Hebb's hypothesis has a molecular basis, although observing exactly how it works has been a challenge to scientists. Given the enormous number of neurons and synapses in the human brain, a better approach of study was to look at very small model organisms. In particular, sea slugs of the genus *Aplysia* have a simple nervous system with a small number of large neurons, like the giant axons in squids that Hodgkin and Huxley worked on decades earlier. *Aplysia* has been used to study the molecular basis of learning, in particular by a group of scientists led by Eric Kandel, who won the Nobel Prize in Physiology or Medicine in 2000. Kandel was able to show that *Aplysia* would learn to avoid an unpleasant stimulus through the remodeling of synaptic connections. In addition to the numbers of synapses between cells, synaptic plasticity also involves changes in

---

[1] As quoted in Norman Doidge's *The Brain That Changes Itself.*

the types of receptors and ion channels present at synapses. In this way, the brain functions like a circuit board on which specific connections, and the relative strength of these connections, can be rearranged over time in order to optimally accomplish important tasks.

Learning and memory also involve connections between neurons that can either increase or decrease the ability of the target cell to create an action potential. In these cases, the cell that is modifying the level of a particular response may not actually be directly involved in the signal's chain of transmission. For example: neuron 1 makes a synapse with neuron 2, and neuron 3 creates an inhibitory connection with neuron 2, so that an action potential transmitted from neuron 1 won't propagate through neuron 2. In this way, neuron 3 actively regulates the transmission of the signal without being directly involved in the propagation of an action potential itself.

Neurons can also form synapses with other types of cells. When we want to pick something up, for example, action potentials that originate in the brain are sent down neurons that connect to our muscle cells via the neuromuscular junction. This is a synapse between a neuron and a muscle cell where the presynaptic neuron releases the neurotransmitter acetylcholine, which binds to receptors on the surface of the muscle cell, initiating muscle contraction. This is the critical component of a "reflex arc," for example, the knee-jerk reflex that takes place when a doctor hits the front of your knee with the little rubber hammer. The hammer blow sends signals via sensory neurons, which activate motor neurons that release acetylcholine to activate your leg muscles. In this case, the reflex does not involve any processing in the brain; the turnaround point between sensing to muscular action is, instead, the spinal cord.

Many of our most critical bodily functions don't require our focused attention. We don't have to consciously tell ourselves to breathe, blink, or digest our food. Even though it involves more than simple reflexes, our autonomic nervous system generally functions without requiring much intellectual contribution. That said, links between the mind and body certainly seem to be real. Stress can give us headaches and stomachaches and may contribute to psychological and medical issues. Relaxation techniques can slow our heart rate and lower our blood pressure. Sometimes a guy can't pee if there is a long line of guys behind him waiting to use the urinal. The relationship between psychology and physiology is extremely complex.

Our consciousness, personality, memory, and ability to solve problems and make complex determinations reside in integrated networks of neurons regulated by mechanisms such as synaptic plasticity and myelination. However, how these neurons are arranged and function together in these highly complex, interdependent roles is currently not very well understood. Two-photon microscopy, which was described in chapter 13, is beginning to let us see deeper into the brain—about 10 mm with subcellular resolution is now possible. However, this requires chemical treatments to remove substances that scatter light. This kind of tissue clearing lets us acquire very high-contrast images from deep in the brain, but not in a living organism. Alternatively, light-sheet imaging, described in chapter 14, gives us the ability to image intact brains in smaller model organisms, such as worms, flies, and fish, but generally not with the resolution necessary to understand what is happening inside individual neurons.

A large group of researchers at the Howard Hughes Medical Institute Janelia Research Campus near Washington, DC, is currently attempting to map all of the synaptic connections in a whole fruit fly brain using electron microscopy (EM). The Fly EM project is made extremely challenging by the fact that conventional electron microscopy images only very small fields of view with very limited depth of field, due to the thin sections required for EM. So this project is rather like stacking up thousands of jigsaw puzzles, each with numerous pieces. A huge amount of data is being collected, and already tremendous advances in sample preparation, EM imaging, and data processing have been achieved in the process of getting this project up and running.

Now is an extremely exciting time for neuroscience. This is an area of major investment by the federal government, as evidenced by the Brain Research through Advancing Innovative Neurotechnologies (BRAIN) Initiative announced by President Barack Obama in 2013. New techniques such as GRASP, GCaMP, and optogenetics, which were described in chapter 17, when combined with technologies such as two-photon and light-sheet microscopy, are allowing unprecedented views into the inner workings of the central nervous system. When combined with network modeling approaches, an integrated understanding of complex neurological processes may be achieved in the coming years based on these types of large-scale collaborations between physicists, engineers, computer scientists, and biologists.

CHAPTER 23

The Immune System: How It Defends Us and Sometimes Attacks Us

○

ORGANS ARE GENERALLY THOUGHT OF AS ISOLATED, SELF-contained structures like the kidney or liver, in which specific cell types work together to accomplish particular tasks. However, some organ systems involve types of cells that are decentralized, working either in groups or as individuals in concert throughout the body. The nervous system, circulatory system, and musculoskeletal system are all examples of distributed networks that carry out complex, nonlinear physiological roles. In these cases, both global and local regulatory mechanisms can be involved. When you hit your funny bone or massage a sore muscle, you feel the effects locally, not throughout your whole body. Individual components of an organ system can be regulated and modified apart from the rest. On the other hand, changes in your heart rate alter the flow of blood through the entire network of arteries and veins—your cardiovascular system. This is an example of a global mode of control over entire organ systems.

The immune system has evolved to become essential in all aspects of health and disease. In addition to protecting us from infection by

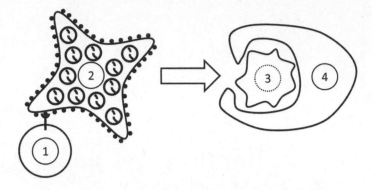

A killer T cell (1) induces apoptosis in an infected cell (2) after recognizing the viral proteins on the cell surface. The residual apoptic (i.e., dead) body (3) is engulfed and devoured by microphages (4).

outside pathogenic agents such as disease-causing viruses and bacteria, the immune system performs critical surveillance; it's constantly on the lookout for cells that have transformed and become cancerous. The immune system can identify and eliminate cancer cells, separating the wheat from the chaff, through alterations in the types and amounts of proteins cancer cells express. In fact, our natural ability to destroy cancer cells is currently being exploited in what is referred to as cancer immunotherapy, which can even include tumor vaccines that educate the immune system to be able to target specific types of cancer cells. However, autoimmune diseases such as rheumatoid arthritis, lupus, and type 1 diabetes, as well as potentially fatal allergic reactions, can erupt due to inappropriate hyper-responsiveness of the immune system, and autoimmune diseases seem to be on the rise. One key to general understanding of the role of the immune system is its ability to discern self from nonself, and its occasional failure to do so. While the ability to adapt to the constant barrage of ever-evolving microbes is key to our survival, our immune system seems balanced precariously between adequate protection and over-reactions that can cause more harm than good.

When functioning properly, the human immune system is rather brilliant in its ability to discern what it should or shouldn't attack and to what degree. When cancer arises, the immune system must be able to identify the newly transformed cells as not truly "self" any longer, but rather as a dangerous aberration, an enemy within. Also, the presence of helpful microbes, particularly in our digestive tracts, must be tolerated. This is generally accomplished by keeping the bacteria inside our intestines and away from other parts of our bodies, while keeping the cells of our immune system in our bloodstream and other organs and tissues on constant alert for undesirable microbes in the wrong place at the wrong time. It should be noted, though, that overreaction of the immune system can be just as damaging as anything that microbes themselves do to us. In fact, since most pathogenic organisms use our cells, organs, and tissues as breeding grounds for their own reproduction, killing the host is generally not in the guest's best interest. So, perversely, just as the immune system weeds out infected cells, it might sometimes act on the individual organism as a line of last defense for an entire population by hastening the death of disease-infected individuals, thereby stemming the spread of the disease.

Pathogens can release toxins or physically impede the normal function of our bodies and cells. Both botulism and tetanus, potentially fatal bacterial diseases, result in symptoms mediated by toxins that inhibit function of our nervous system. That said, much of what we experience when we get ill, such as fever, chills, and a stuffed nose, is actually more the result of the body's *response* to infection than a direct result of any action on the part of the offending microbe. At the most extreme, allergic reactions to completely innocuous substances can result in anaphylactic shock, in which our throats close up and we suffocate; our own immune systems can literally kill us for no good reason.

The human immune system is very complex and has evolved over eons to include several different facets and types of responses. These can be divided into two general categories: innate and adaptive. Our innate defenses are the hardwired, permanent aspects of our physiology that help prevent infection. They include the skin's function as a barrier, the acids in our stomach that digest many unwanted invaders, and the mucus in the respiratory tract that filters out pathogens. The most evolutionarily basic aspect of our innate immune system involves a series of proteins referred to as complement factors. These proteins simply float around in the blood, and when they encounter pathogens, they form a coating around the invader, more or less putting it in jail. This is a very primitive protective mechanism. Many evolutionarily ancient types of multicellular animals such as horseshoe crabs and sea urchins have complement-like systems.

On the other hand, the adaptive immune system has the ability to learn from past exposure to pathogens. Adaptive immunity is why you get chicken pox only once, and it's the key to how vaccines work. It changes over time. Compared to innate protections, adaptive immunity involves a much more complicated and tightly regulated series of events that includes many types of cells that enable the body to respond to infectious agents almost immediately.

Most of our immune cells are lymphocytes, so called because they travel through the lymphatic system, a series of vessels and ducts that connect with and run parallel to the primary circulatory system through which our blood flows. Some lymph cells function as foot soldiers, actively destroying pathogenic organisms and infected cells. These include killer T cells, or more technically, cytotoxic T lymphocytes. (You will remember from chapter 18 the T refers to the thymus gland, where these cells mature.) Another example of a critical frontline immune cell is macrophages, which engulf microbes, infected cells, and fragments of cellular debris—for instance, following a traumatic injury or in clearing away the remnants of a target of a killer T cell. We also have helper T cells, which act like commanders in the field, mobilizing the appropriate response depending on the type and location of a particular microbial threat.

Very similar to helper T cells are regulatory T cells. (If you want to seem very scientifically hip, refer to these as Tregs—two syllables, as in T. rex.) Tregs can globally modulate particular aspects of immune response, sort of like an admiral in charge of a whole fleet. Because Tregs so profoundly affect health and disease, they have recently become the focus of many new drugs—for example, to treat autoimmune diseases. Because these drugs are more precise in modulating immune response, they are a welcome advance beyond the classical immune-suppression drugs historically used to prevent responses such as organ-transplant rejection.

Another type of lymphocyte at the center of much of the activity of the immune system are the B cells, which are derived from our bone marrow. B cells make antibodies. Antibodies are proteins that are either on the surface of a B cell or secreted into the extracellular space. Antibodies are structurally formed of two main parts: the constant domain, which is roughly identical across all specific types, and the variable domain, which differs from one antibody to the next. Antibodies bind things. That is their job, and they do it very well, so that the strength of the interaction between an antibody and its cognate antigen (the toxin or other foreign substance that's attacking us) is very tight. Once an antibody binds its antigen, this "immune complex" serves as a red flag to the rest of the immune system, setting off a series of events that serves to isolate and eliminate the threat. Binding of an antigen—for example, a secreted bacterial toxin—to an antibody on the surface of a B cell will stimulate that B cell to begin producing and releasing more of that antibody. Incredibly, B cells are even able to alter the DNA code of antibodies, making them even more specific and potent. Moreover, the binding of an antibody secreted from a B cell to a pathogen or infected cell triggers killer and helper T cells—the foot soldiers and field commanders—to launch an attack. Thus, the B cell is a bit like Congress, the president, and the Joint Chiefs, identifying an enemy attack, declaring war, and directing the overall effort throughout the campaign.

Once a threat has been eliminated, the B cells stop actively producing and releasing antibodies, but they don't disappear. Rather, they remain in a dormant state as "memory" B cells, ready to spring into action immediately if the same adversary returns. This is the key to how many vaccines work. We are exposed to a generally harmless antigen, either via a heat-killed pathogen or in isolation apart from the rest of the actual microbe. The body then mounts a short-term antibody response to this perceived threat, which leaves us with a new regiment of memory B cells specific to that antigen that will be ready if we're ever attacked by the whole, intact pathogen. This is like having the immune system participate in a training exercise in preparation for a potential future attack.

As we've already noted, the immune system that protects us against invading foreign pathogens can also go haywire and attack our own cells. And autoimmune disease is in fact on the rise. One possible explanation involves the dangerously misaligned connection between two key aspects of the immune system. The first is that adaptive immunity only works if there is something to adapt to—it relies on the presence of pathogens to learn and develop appropriate responses. The second is that the immune system has the potential to do us harm. Autoimmunity is a bit like friendly fire. In the absence of appropriate training, our immune soldiers are losing the ability to discern friend from foe. This is most likely happening, at least in part, due to something that otherwise seems a huge benefit to our health: the relatively sterile environments we all live in nowadays.

Prior to the twentieth century, life was pretty squalid. People generally didn't wash their hands with antibacterial soap after using the toilet or before they ate. Genteel Victorians certainly valued cleanliness and were fastidious about their appearance and would be ashamed to show up as dinner guests with mud on their cuffs and hems. However, the concept of microscopic pathogens didn't exist until the late 1800s, when Louis Pasteur proposed the "germ theory" of disease. Have we now gone too far in the other direction? Are the filtered, purified water we drink and the hand sanitizer we use isolating us from the microbial world to too great an extent? If our immune systems don't get the exercise and training they need, will they not function as well as they should?

If our immune systems are honed and shaped by encountering microbes in our daily lives, what happens as we sanitize our existences? Eventually, we are going to fly on an airplane, use a public toilet, touch a doorknob, or kiss a child who has just come home from school. Will our bodies be ready for that?

Furthermore, without the normal microbial encounters to train our immune system and thus regulate its responses, things can go completely off the rails. Recent work is focusing on the role of the bacteria in our digestive tracts, our "gut microbiome," in health and disease. The rise in popularity of so-called "pro-biotics" that contain live bacterial cultures to assist in digestion, and with other potentially beneficial effects, demonstrates the alterations in the gut microbiome many in the industrialized world seem to be facing. In fact, recent evidence suggests that oral antibiotics can have lasting negative effects on the bacterial species that we need in our digestive tracts. As our diet becomes more limited and full of processed food, and as we reduce our exposure to the microbes with which humans have long coexisted, what is this doing to our ability to digest food, to our general health and well-being, and to the ability of our immune systems to function?

PART SIX

THE PROFESSION
OF CELL BIOLOGY—
THE GOOD, THE BAD,
AND THE FUTURE

CHAPTER 24

How We Almost Cured Cancer

◉

I N 1971, JUDAH FOLKMAN PUBLISHED AN ABSOLUTELY GROUND-breaking and revolutionary paper in the *New England Journal of Medicine* that outlined an entirely new hypothesis regarding the mechanisms that control cancer progression. Folkman was a surgeon, and like many other physicians, he had long noticed that tumors, especially the larger and more aggressive ones, were often found intimately associated with the patient's blood vessels. A tumor is a solid mass of rapidly growing cells that requires oxygen and nutrients, as well as a way for waste products to be removed. This means a nearby vascular supply of blood is necessary for tumor growth and cancer progression. The prevailing wisdom was that only those tumors that start forming near arteries and veins would survive and grow. What Folkman suggested, however, was that tumor cells actively released substances that would *attract* nearby blood vessels to move closer, facilitating the growth of the tumor.

Angiogenesis is the process that governs growth of blood vessels. The cells that make up the surfaces of arteries and veins, which are called endothelial cells, start to multiply and then actively migrate—crawl—toward the site where more blood is needed, where they coalesce into tubelike vascular structures. Angiogenesis occurs primarily during development, especially in utero, when new organs

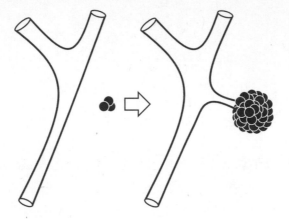

Angiogenesis provides a blood supply that promotes
tumor growth.

and tissues are growing and increasing the need for blood supply.
Angiogenesis also takes place during tissue repair—for example, in
wound healing or in response to a bone fracture. In the case of wound
healing, cells at the site of damage will release factors into the extracellular space surrounding the wound that attract endothelial cells, as
well as immune cells, to move in that direction. Folkman proposed
that tumors released chemoattractants and other angiogenic factors,
and that this would stimulate angiogenesis at the site of the tumor so
that the cancer cells could grow.

One further aspect of Folkman's later work was even more fantastic. He proposed that cancer cells were also able to release other
anti-angiogenic factors that would prevent the growth of other nearby
tumors, creating a sort of arms race in which different tumors com-
peted for the limited blood supply in a particular region of the body.
The general idea was that a successfully vascularized larger tumor
would release anti-angiogenic factors into the surrounding vicinity
to prevent smaller, competing tumors from recruiting blood vessels.
What led Folkman to this conclusion was the observation that after
a surgeon removed a large tumor from a patient, there would often

be a rapid and immediate growth of other microscopic tumors that seemed to have been waiting in the wings, invisible to the surgeon.

The important consideration here is that tumors can only grow to a small size, a few millimeters, without any need for a direct blood supply. The substances that cancer cells need to take in and get rid of can simply float across small distances, but when the tumor gets bigger than this critical size limit, about the size of the head of a pin, it must recruit new blood vessels if it is going to survive. So Folkman's idea was that these tiny micro-tumors were already present but too small to be seen by surgeons. As they were awash in anti-angiogenic factors produced by the nearby large tumor, they were unable to use their own small supplies of pro-angiogenic factors to any avail. But once the source of this repression, the large tumor, was removed by the surgeon, the tiny tumors would be free to induce their own increased local blood supply and grow rapidly.

Most importantly, Folkman surmised that if these anti-angiogenic factors existed, they could be purified, identified, and employed in the fight *against* cancer, turning the tumor's own competitive advantage against itself. This led to a long and complicated series of studies that lasted several years and required Folkman and his team to develop innovative new experimental techniques. At the end of it all, Folkman published his conclusions that several anti-angiogenic factors did exist and they could be used in the fight against cancer. Two anti-angiogenic factors in particular, angiostatin and endostatin, were seen by big pharmaceutical companies as potential magic bullets in the fight against cancer.

Thus, anti-angiogenic factors were thought to represent an entirely new way to treat solid tumors. This excitement led Jim Watson to reportedly state, "Judah is going to cure cancer in two years," as quoted by the *New York Times* in 1998. (However, it should be noted that subsequent reports state that Watson denied ever have used these exact words.) At that time, the results of studies testing angiostatin and endostatin as cancer treatments in humans were eagerly anticipated.

In fact, the potential lifesaving effects of these much-hyped miracle drugs were so enticing that over a thousand cancer patients sought to be included among the handful of subjects who were given endostatin in its first clinical trial.

Obviously, Watson was wrong. The clinical use of anti-angiogenic factors in most cancer patients has been disappointing. Sadly, Folkman died of a heart attack in 2008, further setting back the field he created.

So, why haven't anti-angiogenic therapies resulted in paradigm-shifting treatments? The answer is not entirely clear, and could at least in part be accounted for by the difficulties often found in translating basic science observations from cell culture or animal models into actual humans with real diseases. However, it also seems to have to do with the observation that, upon close inspection, the blood vessels that form at tumor sites are actually very disorganized. Although the tumor is able to promote angiogenesis, it often occurs in a very haphazard way; rather than looking like efficient conduits for blood flow, the arteries and veins associated with the tumor look like a tangled knot of spaghetti. So it seems that when treated with substances like angiostatin and endostatin, the blood vessels in the vicinity of the tumor *do* decrease in total number, but what you are left with may actually be *more* efficient at delivering oxygen and nutrients and removing waste products. In other words, it seems that anti-angiogenesis therapies don't actually choke off the blood supply to tumors as hoped, and, worse yet, they can normalize the vascu-larization of tumors, *increasing* local blood flow. Recently, however, this seeming failure has actually been used to the advantage of the physician and patient through the combination of anti-angiogenesis agents with chemotherapy.

You may already have guessed why this is so. The conventional treatment of cancer generally involves introducing cytotoxic chemi-cal agents that kill rapidly dividing cells. In a healthy adult, most cells are not actively dividing. Only a few select types of cells, such as those of the immune system and in hair follicles, regularly undergo replica-tion in great numbers. This is why chemotherapy decreases immune system function and causes patients' hair to fall out. Researchers realized that putting anti-angiogenisis factors together with conventional chemotherapy would improve blood flow to the tumor, resulting in greater accumulation of the cytotoxic agents in the tumor and better results for the patient. So, in the end, anti-angiogenesis factors do seem to be making a significant impact against cancer, even if it is somewhat contrary to the way Folkman and Watson predicted.

To paraphrase Robert Burns, the best-laid plans of mice, men, and scientists often go awry, but sometimes that produces good results and new discoveries that no one expected. Other times, unexpected results can be evidence of something potentially sinister.

CHAPTER 25

Ethics, Ambition, and the Greatest Discovery That Wasn't

T HE PROFESSION OF ACADEMIC SCIENCE HAS A STRONG ETH-
ical code and maintains strict standards of evidence and
peer review. Scientists also rely upon the repetition of
results to confirm new discoveries. The word "science" is derived from
the Latin word for knowledge, and knowledge—the truth—is ideally
our only real goal.

On the other hand, scientists are prone to human error, ranging
from simple mistakes or misinterpretation to the desire for fame and
fortune, ensuring promotion up the academic ladder, or even just not
getting fired. Recently, a story unfolded of a monumental "discovery"
and how it was quickly debunked, with sad consequences for a number of people and institutions.

It's also a story of one of the most exciting and promising scientific frontiers—the healing potential of stem cells—so before we get
to the juicy human drama, we need to talk a little science.

Like the different cars on the road, cells are defined by specific
distinguishing structures and functions. A Chevy pickup truck and a
Cadillac limousine are made by the same company with pretty much
the same materials, but we can immediately recognize them as different vehicles designed for different purposes. Similarly, a muscle cell

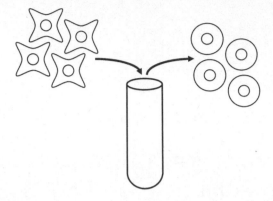

Haruko Obokata claimed that simply by placing mature differentiated cells into a mildly acidic solution, they could be converted into stem cells. This was quickly proven to be false, but whether or not Obokata made an honest mistake remains a mystery.

and a nerve cell exhibit distinct characteristics. Cells have particular roles that arise from the specific components they contain.

Just as a lump of metal can be forged into a sword or a plowshare, all the cells in your body started out with the same DNA in the nucleus—an identical genome. However, from this common origin can spring any of the hundreds of specific cell types found in the adult body. This is mostly accomplished by regulation of which genes get expressed, as well as which particular splice variants of those gene products the cell makes into protein. Proteins create the structures and components that allow a cell to fulfill specific roles and perform particular tasks. However, what about the initial lump of metal? That is a stem cell.

A stem cell is defined by two characteristics. First, it is not differentiated, that is, it hasn't become specialized as, for example, a muscle or bone or blood cell, although it has the potential to do so. Second, it actively replicates. Differentiation takes place mostly during embryonic development, as cells adopt a final, mature identity. Stem cells have the potential to differentiate into any and all of the many types of mature cells; they are the undifferentiated forms from which mature tissues and organs form. When skin cells become skin cells and kidney cells become kidney cells, these differentiated cells arise from other less differentiated forms, starting with stem cells, and the process is generally thought to be unidirectional. Normally, cells in the body will not "de-differentiate"—they will not revert to less specialized forms. Mature cells are "terminally differentiated." Once a neuron, always a neuron. However, it seems that scientists have developed ways to actively reverse the process of differentiation in the lab, essentially making stem cells from adult cells.

The therapeutic potential for using stem cells is extremely enticing. Many deadly and debilitating human diseases arise from the loss of specific cell types. Ranging from recovery from a heart attack to treatment for things like Alzheimer's disease, the ability to introduce stem cells that would replace lost, damaged, or dead differentiated cells would be revolutionary. Furthermore, experiments have demonstrated that environmental signals—for example, factors released from neighboring differentiated cells in a tissue—can stimulate differentiation. Imagine that you could inject stem cells into a damaged tissue and the local signals would automatically guide repair on a cellular level. This kind of stem cell therapy is a very real possibility.

The problem is, where do you get the stem cells? Unless you are going to generate your own (more on that later), there are two choices, two different types of stem cells: embryonic and adult. Embryonic stem (ES) cells are found during a very early stage of development. They are formed about a week or two following fertilization and are simply a small mass of completely undifferentiated cells. ES cells have the potential to form all the different cell types normally found in an adult. However, given where they come from, getting them and using them is problematic. In addition to practical and ethical concerns, such as the use of fertilized human embryos for research, there are also factors that limit the therapeutic potential of ES cells.

As with any foreign tissue, injecting or implanting ES cells from one individual into another could result in significant problems due to rejection, as the immune system is constantly on the lookout for foreign invaders that register as "nonself," including cells from another human. Implementation of ES cell therapy would need to take this into consideration. As ES cells come from a very early stage of embryonic development, it would be too late to get our own, lacking a time machine. So without some sort of genetic engineering of the cells to reduce the chances of rejection, or the use of drugs that would inhibit the recipient's immune system—a potentially dangerous option—use of ES cells for tissue repair is currently not very feasible.

So what about adult stem cells? Most adult cells that are in a state of terminal differentiation do not undergo active replication and can stay alive and in place for long periods of time. However, some cell types found in adults are not fully differentiated and do continue to undergo cell division. Skin cells, cells in our hair follicles, many blood cells, and cells lining the digestive tract must be constantly replaced due to loss during normal daily life. These come from pools of stem cells that reside within those tissues throughout adult life. Generally, an adult stem cell is not truly a blank slate like an ES cell. It seems adult stem cells from one organ do not contain the potential to differentiate into any and all cells in the adult. Rather, adult stem cells seem to be somewhat tissue-specific. The stem cells in adult bone marrow generate blood cells, but these hematopoietic stem cells are of a different type from those stem cells that allow our skin to heal without a scar if we cut our finger while slicing onions.

Medical treatment employing adult stem cells has actually been around for decades in the form of bone marrow transplants. Radiation and chemotherapy during cancer treatment kill any dividing cells in the body, including hematopoietic stem cells. So the answer is to transplant bone marrow, which is rich in hematopoietic stem cells, from a donor into the cancer patient. This is stem cell therapy, and it has been a medical reality since the 1950s.

However, hematopoietic stem cells are not a panacea because they cannot differentiate into all other types of cells. There are other adult stem cell pools that are specific to the tissues in which they are found and limited in the cell types into which they can differentiate. In the case of skin, the presence of adult stem cells led to another stem cell therapy that has existed for decades. Developed by Howard Green, a physician-scientist at Harvard Medical School, as a means of treating patients with severe burns, the therapy involves removing small pieces of healthy skin from the burn victim that are then cultured in the laboratory. After these patches have increased in size, they are grafted onto the burned areas. This therapy completely relies on the presence of skin stem cells and has saved many lives.

While it is well documented that tissues in which cells turn over rapidly, like the skin, contain relatively large numbers of adult stem cells, it is also likely that many other types of adult stem cells exist that can help recovery in times of damage or disease. However, for many organs, such as the heart, lungs, kidneys, and brain, insufficient quantities of tissue-specific stem cells can lead to debilitating disease. Although it might be possible to find a few adult stem cells in a damaged or diseased tissue and stimulate them to replicate themselves, it would be much easier to take normal, terminally differentiated cells and turn them into stem cells. While it might sound like the stuff of science fiction, in 2006 a Japanese researcher named Shinya Yamanaka did just that. By introducing four different proteins that stimulate transcription of genes associated with stem cells, Yamanaka created induced pluripotent stem (iPS) cells. Like ES cells, iPS cells seem to have the potential, or "potency," to develop into cells of all different types. Yamanaka was awarded the Nobel Prize in Physiology or Medicine in 2012 for his work.

Unfortunately, many of the techniques currently available for generating iPS cells are complicated and potentially unsuitable for therapeutic use. One of the proteins Yamanaka used to generate iPS cells, the MYC oncogene, is strongly associated with the development of many types of cancer. Thus, methods for generating iPS-like cells that would be more easily and safely developed into therapies are being very actively pursued. In January of 2014, into this space stepped Haruko Obokata, a young Japanese researcher less than a decade out of college. The powerful spotlight placed on Obokata's work would very rapidly reveal a strange, confusing, and ultimately tragic tale of scientific misconduct.

Obokata published two papers in the January 30, 2014, issue of the prestigious journal *Nature*. To show you what a big deal that was requires a bit of digression into the field of scientific publishing.

Publishing individual research reports in scientific journals is the main currency of science. For the vast majority of biological scientists, three journals sit at the top of the heap: *Nature*, *Science*, and *Cell*. It is widely thought that a paper in one of those journals can make a scientist's career, and a string of such papers can cement an author among the luminaries in his or her field. This is especially true if the researcher in question is the first (lead) or last (senior) author listed on a paper. Generally, the first author is the one who did the most work, and the last author is the leader of the research team. In fields where papers might have as many as ten or fifteen authors, the first and last positions on the author list matter disproportionately. Obokata was listed as first author on both of the *Nature* papers in question, earning her both the lion's share of the fame and, ultimately, shame.

Journals are ranked by a measure referred to as impact factor. Papers in the best journals are referred to as high impact. The impact factor is a measure of how often, on average, papers in that journal get cited in other published work, acknowledgment that the paper made a significant impact on the field. Many historically respected journals have impact factors in the range of 5 to 10. Although papers with the highest impact can be cited hundreds or even thousands of times, many papers only get little or no mention in other publications. Of course, it takes time to accrue citations, and impact factors are lower for newer journals and those that publish the most papers or have less stringent criteria for acceptance. However, older and historically important journals can also drop in impact factor when they change their editorial criteria or publication strategy—or when the specific field they cover is no longer "hot."

There is, for example, a paper from the laboratory of Roger Tsien, who went on to win the Nobel Prize for his work on GFP and other fluorescent proteins, titled "A new generation of Ca2+ [calcium] indicators with greatly improved fluorescence properties." This paper was published in the *Journal of Biological Chemistry* in 1985 and has been cited in other papers more than 20,000 times. However, since papers in that journal are not very often cited, overall the *Journal of Biological Chemistry* currently has an impact factor of only 4.5. *Nature*, on the other hand, has an impact factor of 41.5, nearly ten times higher.

Back to Obokata. What did she report that was so amazing as to result in *two* papers in a single issue of *Nature*? Nothing short of a miraculously simple technique for creating stem cells. According to Obokata (and her coauthors from Harvard and Japan), stem cells could be generated by exposing differentiated cells to specific stresses under particular controlled laboratory conditions. These were called stimulus-triggered acquisition of pluripotency (STAP) cells. Obokata's papers described the techniques used to make STAP cells, such as incubating certain differentiated cells from the spleen, the organ involved in blood-cell formation and turnover, in a weakly acidic solution. Together, the papers seemed to prove that following exposure to these kinds of stressful conditions, some cells would appear to revert to a stem-cell-like state. They also demonstrated that STAP cells could differentiate in a living animal into every kind of tissue. To the field of scientists who, only a few years before, had seen a Nobel Prize awarded to Yamanaka for a much more complicated and less efficient technique to generate stem cells with limited therapeutic potential, this seemed the stuff of fantasy.

Obokata was only about twenty-five years old when she started her stem cell work, and it is still not entirely clear how ultimately erroneous results with such groundbreaking implications could have slipped through all the checkpoints and into print. However, it is now clear that Obokata's papers reported false claims. The papers have been retracted, removed from the published record, and Obokata no longer holds the group leader position that this work earned her at the prestigious Riken Center for Developmental Biology in Kobe, Japan.

One optimistic perspective on the Obokata scandal is that it provides a window into the potential benefits of the Internet era to science. Science has always been self-correcting. If a paper's conclusions are proven by subsequent work to be incorrect or incomplete, the field as a whole moves forward. Of course, fraud and malfeasance have always existed in science as in all other fields. But now that the Internet allows information to be shared so quickly between so many people in such broad conversation, shoddy or false research findings tend to be exposed quickly. Almost immediately after Obokata's papers came out, labs around the world tried to generate their own STAP cells. So far, none has succeeded. Within a few months, Kenneth Lee's group from the Chinese University of Hong Kong published a paper in the open-access journal *F1000Research*, after *Nature* reportedly refused to publish it, saying that they were unable to make their own STAP cells using Obokata's procedures. As a result of an internal investigation at Riken, a team of scientists that included Obokata tried for several months to reproduce the findings from the papers and were also completely unsuccessful. *Nature* did eventually publish a paper in September of 2015 in which seven independent labs in the US, Israel, and China reportedly failed in a total of 133 attempts to make STAP cells. Thus, it seems that STAP cells never in fact existed.

Further analysis of Obokata's work has raised a series of red flags. Why these were not identified earlier, before publication, is a good question. A cornerstone of Obokata's results was the observation that when stressed, the cells started to express a marker protein associated with stem cells. This was determined by looking for green fluorescence, as the stem cell marker was tagged with GFP. Unfortunately, however, cells often exhibit autofluorescence, generating on their own a dim glow, especially in the green channel, and in particular when stressed, dying, or dead. Subsequent analyses have demonstrated that the green glow in Obokata's cells wasn't because they were expressing the GFP-tagged stem cell marker but was merely autofluorescence.

Many other issues with Obokata's work have been uncovered. One of the figures from her papers appeared to show evidence of image manipulation. (Authors can fabricate the results they want to present by cropping out parts of an image, altering the contrast or shading in dishonest ways, or combining bits from different images into one to demonstrate something not otherwise evident.) Elsewhere, two figures purportedly depicting different results appear to contain the exact same image. Similarly, in a different figure it appears that an image was reused from Obokata's PhD thesis. Without proper attribution, this would "only" be self-plagiarism. However, in this case the image was apparently used to depict the results of two completely independent experiments, one on exposing cells to acid and another following application of physical pressure on cells. Further, it was reported that part of a section that described some of the analytical methods used was plagiarized, basically cut and pasted from another article without attribution. Finally, subsequent genetic analysis of the STAP cells Obokata supposedly created revealed the presence of contamination from embryonic stem cells derived from a mouse line with a distinct genetic background. Whether this might have been the result of deliberate misconduct or simple confusion, or even sabotage, is not clear.

It is still not known to what extent Obokata's actions were intentionally meant to deceive. She has not been very forthcoming with the press, and what she has said in response to the evidence against her boils down to claims of inexperience, disorganization, and poor mentorship and oversight. She asserts that she did not mean to falsify or manipulate the evidence presented in her *Nature* papers, and that her science was sloppy, not dishonest.

One take-home message from this affair is that greater oversight is required both of trainees in the lab by their supervisors and by editors of journals. The section of Riken where Obokata worked has been completely reorganized, and *Nature* has provided assurances that this type of scandal cannot occur again. It is laudable that this scandal was resolved before significant investment by governments or biotechnology and pharmaceutical companies might have been misallocated, and not to mention before any potential STAP-cell therapies might have been tested. Nonetheless, significant damage was done.

Scientific fraud erodes the trust and respect of the public, jeopardizing support and funding. Obviously, Obokata's career and reputation are in tatters, her PhD was even rescinded, and several of her colleagues and coauthors have emerged tainted from this affair. Saddest of all, Yoshiki Sasai, one of Obokata's supervisors and the senior author on one of the STAP papers, hanged himself in the stairwell of Riken in August of 2014, about a month after the two *Nature* papers were retracted. It was reported that before committing suicide, Sasai wrote a note to Obokata imploring her to "be sure to reproduce STAP cells." Thus, it would seem that even following retraction of the paper and failures to replicate the results, at least Sasai still believed in the potential of the STAP phenomenon. What Obokata was thinking, we can only imagine.

CHAPTER 26

The Nano Revolution

◉

W ITH RECENT DEVELOPMENTS IN SUPER-RESOLUTION
microscopy, also known as nanoscopy, we are now
capable for the first time of truly visualizing many
of the structures, compartments, and processes inside of living cells
at their true scale and context. Life unfolds at the nanoscale, and
we now stand at the border of a new world where things are no
longer obscured from our view by the resolution limit of the light
microscope. With newly developed super-resolution fluorescence
"nanoscopes," we can now simultaneously image multiple different
molecules, resolving them from each other in the busy and confusing context of a living cell. Where once we were blind, now we can
see. With the recently developed tools for labeling and imaging cells,
even in a living model animal, we can finally visualize life at the subcellular scale.

In addition to these groundbreaking advances in imaging, the
emerging field of nanotechnology is giving us the tools to label, track,
and manipulate cells and cellular processes at the finest levels of scale
imaginable. The production of materials and tools, even whole nano-
machines, out of tiny assemblages of precisely arranged molecules is
providing for the first time the ability to measure, alter, and influence life at the nano scale. We are entering an entirely new era of
science in which the conventional rules of mechanics and chemistry

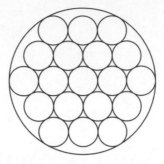

Smaller structures have a greater combined surface area
than a large structure of the same total volume. Very tiny
things, nanoparticles, have so high a ratio of surface area
to volume that they behave much differently than larger
objects, and in unpredictable ways.

are superseded by quantum effects. Cellular interactions at the level of
the nano-bio interface cannot easily be predicted, but will ultimately
be modulated and harnessed to provide astounding breakthroughs in
disease diagnosis and therapy.

We're seeing astonishing inventions—two-dimensional pla-
nar arrays of atoms such as graphene (a sheet of carbon one atom
thick) and extremely long and thin nanowires—that hold tremendous promise for computational and electrical applications. But at
the center of much current biomedical work are nanoparticles. Technically defined as symmetrical structures less than 100 nm across,
about 1,000 times smaller than a human hair, nanoparticles have the
potential to revolutionize biology and medicine. Nanoparticles have
always existed in nature; they are present in clay and produced in fires
from the breakdown of wood and organic matter into tiny clusters
of organic compounds and metals. However, humans have recently
begin to engineer, produce, and manipulate nanoparticles for potential use in numerous biomedical and industrial applications.

What makes nanoparticles so fascinating and powerful also
makes them potentially dangerous. Materials this small tend to

exhibit properties that are quite different from the same substances at the normal macro or bulk scale. At the nano scale, things like electrical conductivity, optical properties, and the ability to participate in chemical reactions change in seemingly unpredictable ways. One major cause of the amazing behaviors of nanoparticles is their extremely high ratio of surface area to volume. The smaller an object is, the more of it is on the surface. Put another way, as particles get smaller, their volume shrinks much more rapidly than their surface area. That is because volume is a product of the cube of the particle's radius, but surface area only depends on the square of the radius.

Specifically, the volume of a sphere is $\frac{4}{3}\pi r^3$, and the surface area of a sphere is $4\pi r^2$. Let's say a basketball has a diameter of 10 inches and a radius of 5 inches. Thus, its volume is about 524 cubic inches, and the surface area is approximately 314 square inches. This gives you a surface area to volume ratio of 314:524, or just about 0.6.

A golf ball is 1.7 inches in diameter, so its radius is 0.85 inches. Ergo, the volume of a golf ball is about 2.5 cubic inches, and its surface area is about 9 square inches. This is a ratio of 9:2.5 or 3.6.

Imagine that you wanted to paint the same volume of golf balls as taken up by a single basketball. The volume of a basketball is over 200 times greater than that of a golf ball (524÷2.5). However, the surface area of a golf ball is only about 35 times less than a basketball (314÷9). That means it would take approximately six times as much paint (200÷35) to cover the same volume of golf balls as a single basketball.

What this means is that lots of tiny little nanoparticles have much more exposed surface area than the same amount of material in one large lump. This increases the potential for binding and interacting with things. Furthermore, the vesicles involved in internalizing material into cells via endocytosis are just about 100 nm across. This means that while larger particles might not be able to easily gain entry into cells, nanoparticles can hitch a ride during endocytosis and get themselves inside of cells. What they do once they get in there is extremely hard to predict, as are many other interactions between nanoparticles and the biological world. A great deal of research is underway in the attempt to understand what happens when particles this tiny interact with living systems. However, the more we learn, the more it seems that each particular case, each interaction between a nanoparticle and a cell of interest, needs to be individually assessed. Quite simply, things at the nano scale are very unpredictable.

Nanoparticles differ in composition, for example, gold versus silver; gold nanoparticles are used for their electrochemical and optical properties while silver nanoparticles are used as antimicrobial agents. Nanoparticles can also differ in size: they are in the range of 1 to 100 nm in diameter. They can be spherical, oval, or even cubic. Their surfaces can be rough or smooth and differ in many other ways. Each of these specific features can affect the function, behaviors, and interactions of nanoparticles with biological systems. Imagine swallowing a wedding ring. It would probably come out the other end, no problem except for the retrieval. Now imagine that you ingested the same amount of gold in the form of tiny nanoparticles. Who knows what would happen or where they would end up.

There are two major areas of interest and inquiry at the intersection of biomedical research and nanotechnology. The first addresses the question of what might happen if nanoparticles created for industrial purposes inadvertently came into contact with biological systems. Silver nanoparticles are added to our socks to keep them from smelling, as they prevent bacterial growth. Nanoparticles made of cerium dioxide can be used to increase fuel efficiency. It has been shown pretty convincingly that our skin is an effective barrier to nanoparticles entering our bodies, and we are well protected from having them leach into our bloodstream. On the other hand, what happens to the silver nanoparticles from your pleasant-smelling socks that end up in the wastewater from your washing machine? Do they end up in lakes, rivers, and groundwater? Are they taken up into fish that we then might eat? How about the cerium oxide nanoparticles that come out in bus exhaust? Do they get into our lung cells if we inhale them? These are but two of numerous examples of how industrial nanoparticles might enter the environment and potentially our cells.

The current general regulatory perspective is that if a bulk material isn't toxic, nano versions of it are considered safe unless proven otherwise. So far, no clear examples have been demonstrated of unintended cell death or DNA damage due to nanoparticles, at least with those currently employed by industry at the concentrations to which we are likely to be exposed. However, there is still a great deal of work to do in this area. Given the unpredictable nature of nanoparticle behavior, the field of nanotoxicology is a very long way from understanding even what kinds of questions to ask and how to go about addressing them. Environmental nanotoxicology is a very active area of research, and scientists around the world are trying to assess different nanoparticles' potential for toxicity and, if possible, synthesize this into a framework for risk assessment that could be applied in a predictive way.

The other main area of inquiry regarding nanoparticles and cells is nanomedicine. There are huge hopes for the integration of nanotechnology into medicine, ranging from imaging and diagnostic agents to therapeutic approaches and medical devices. Super-paramagnetic iron oxide nanoparticles, or SPIONs, have been approved for clinical use, although their application is not currently widespread. SPIONs, as the name suggests, are magnetic nanoparticles. That means that they can be used in magnetic resonance imaging (MRI) studies. MRI works by converting information about the magnetic spin of protons in hydrogen atoms in the body's tissues into images that help doctors diagnose injuries and diseases. MRI can be used to study soft-tissue injury, such as torn cartilage in your knee, or to identify tumors in cancer patients.

The addition of SPIONs decreases the signal detected by the MRI, causing the areas where they are present to appear dark in an MRI scan. This can be extremely useful, since normal tissue will readily take up SPIONs while tumors generally will not. In this way SPIONs act as a contrast agent, making it significantly easier for doctors to visualize the cancer. Thus, SPIONs are a great example of nanotechnology applied to a biomedical issue. Or at least they might be. It has not really been proven that SPIONs perform significantly better than conventional MRI contrast agents, so they are not often employed in clinical settings.

Much of the excitement over nanomedicine is in regard to potential applications. The ideal is a magic-bullet approach in which multifunctional nanoparticles could be used to diagnose and treat diseases in ways that are more potent and specific than current therapeutic options. The hope is that nanoparticles can be imaged in the body with some kind of label, for example, a fluorescent or luminescent tag. These nanoparticles would also need to have some way to target to a specific cell type of interest. This could be accomplished by having something bound to their surface that would allow them to stick selectively to target cells—perhaps by binding to a particular receptor over-expressed on the surface of cancer cells. Finally, the nanoparticle would have some type of payload to deliver. This could be a drug—for example, a powerful painkiller targeted to the site of an injury—or a potent cytotoxic agent that would only kill cancer cells.

The problem is that every time you alter any component of a nanoparticle by sticking labels or drugs onto it, you run the risk of drastically changing its behavior—for example, how well it stays in solution, what it binds to, and how it interacts with cells. Thus, while many aspects of the idealized scenario have been realized in research laboratories, actually getting it all to work together as planned in a living human is another story. Development of new therapies is, and should be, a slow, deliberate, careful process, and there has been an explosion of new nanomedicine research projects. These explorations are headed in so many different directions—looking for solutions to specific problems using different types of nanoparticles, labels, targets, and therapeutic agents—that the field has become fragmented.

We have a lot of irons in the fire but have yet to forge truly groundbreaking developments. This new technology remains both tantalizing and frustratingly unpredictable.

So far, the state of nanotechnology is important mostly for its potential. While we've yet to gain much power to diagnose and treat diseases through its application, there is this good news: it seems unlikely that swarms of out-of-control nanobots will digest all of nature into gray goo—an actual scientific term coined by nanotechnologist Eric Drexler in the mid-1980s. The risk of invisible autonomous agents wreaking havoc was also highlighted in Michael Crichton's 2002 science fiction novel *Prey*, and similar concerns led Prince Charles in 2003 to request a Royal Society report on potential risks of nanotechnology. Over a decade later, nanotechnology has advanced considerably, but not in ways that greatly threaten or improve our health and well-being. Zinc oxide nanoparticles in sunscreen and incorporation of nanotechnology into electronic circuits certainly represent important advances, but we still seem to be standing at the edge of a revolution. Whether this is due to regulatory red tape, limited potential of nanotechnology to change the world, or a lack of creative ingenuity, only time will tell. Historically, from the electric lightbulb to the Internet, it has taken decades for new technologies to move into widespread, general use. Even when basic research in the lab advances rapidly, application to commercially viable products and disruptive technologies can take much longer.

CHAPTER 27

Money, Power, Ambition, and the Pursuit of Knowledge: Thoughts on the Current State of Science

W E ARE LIVING IN A GOLDEN AGE OF BIOMEDICAL SCIENCE, with dazzling advances reported almost daily. We now have the power to rapidly sequence and manipulate whole genomes and the ability to visualize events at the molecular scale in live cells. Global interdisciplinary networks of scientists are combining computational modeling with technological advances that give us entirely new perspectives on critical questions for human health and disease. I hope you feel that this book has been useful in clearly describing some of the more important and interesting breakthroughs in biomedical research science, particularly from the point of view of cell biology and microscopy. If you feel inspired to learn more or even train to become a scientist yourself, there are some important final questions to consider.

First of all, how does one become a scientist?

In the Victorian era, what you most needed was money—your own or that of a patron—time and space for your work, a suitable mentor, and access to a well-stocked library. Our networks of university labs,

research institutes, and pharmaceutical or biotech companies, popu-
lated by a complex hierarchy of trainees, supervisors, and principal
investigators, or PIs, are a relatively recent development. That said,
science still widely employs a mentorship-apprenticeship model. The
only real way to learn science is to do science. You can't read a cook-
book and then walk into a professional kitchen and turn out per-
fect gourmet meals. Scientific training starts with basic fundamental
techniques, the equivalents of chopping onions or stirring a pot of
pasta sauce, then moves on to the more advanced skills and techniques, up to the point of running one's own lab.

The PI is the chef. In larger labs, the PI generally doesn't spend
much time doing the experiment or training the new lab members
on basic techniques. Wolfgang Puck and Daniel Boulud don't teach
people how to chop onions. A hierarchical network extends from the
PI to the senior lab members all the way down to the undergraduate or even high-school-project student. Although medical doctors, MDs, also known as "real doctors," at least by my grandmother,
are certainly scientists, and many MDs conduct excellent research,
the Doctor of Philosophy, PhD, is generally the degree that most
card-carrying scientists achieve. Nearly all PIs of research labs that
focus on areas like cellular and molecular biology have PhDs.

There are three general types of biomedical research: basic,
translational, and clinical. Clinical research is the easiest to understand, and often the most complex to perform. It entails testing of
new drugs, devices, and procedures on humans directly aimed at
diagnostic or therapeutic advancements, usually culminating in a
clinical trial. The burden of proof over the conventionally accepted
therapeutic standard is often extremely high, and the ethical and
statistical hurdles to the researcher are very complicated. It often
takes huge teams of researchers, administrators, doctors (both "real
ones" and PhDs), and nurses to conduct clinical research, particularly with patients in hospital settings. Although the pathway for
the development and ultimate deployment of a new drug can seem

onerous, the process involves numerous safeguards and a suitably high burden of proof.

At the other end of the spectrum, basic science is the bedrock of biomedical research. This is where the fundamental understanding of natural phenomena is achieved. Although basic scientists are often working toward some general goal related to human health, this usually takes the form of increasing our knowledge on a topic rather than trying to find a specific solution to a particular clinical problem. Individual basic scientists usually talk about wanting to understand cancer or Alzheimer's, rather than hoping to cure them. The promise of basic research is that, ultimately, through the combined efforts of many researchers working together in a loosely defined network, diseases will be understood well enough that the translational and clinical researchers can directly improve diagnosis and therapy. Furthermore, this often involves complex indirect and direct relationships between academic researchers and pharmaceutical and biotech companies.

Nearly all of the work that wins Nobel Prizes, as I have highlighted in this book, starts out as basic research. When Osamu Shimomura decided to figure out what makes a jellyfish glow green, he was probably not anticipating a future where fluorescent tumor cells could be tracked in a living animal to test cancer therapies. However, times are changing; more and more, biomedical research requires a direct beneficial application to be deemed worthy of funding. This focus on applied research certainly has many benefits, but not all advances can be predicted.

Translational research exists to convert basic understanding into clinical or industrial application. It can be directed toward developments in any area of medicine, surgery, or diagnosis, or even economic or societal advancements. Translational research can refer to interdisciplinary teams of MDs and PhDs working together or can be the focus of individual researchers, such as those who hold both MDs and PhDs.

Medical school involves courses in anatomy, physiology, diagnostics, and many subspecialties followed by a gradual development of hands-on experience. Generally speaking, the process of getting a PhD is much more open-ended, as well as significantly more focused. In most biomedical graduate schools, at least in the United States, there is no firm graduation date set when you begin your PhD studies. Rather, you are permitted to write and ultimately defend your thesis whenever a committee of faculty members, usually not directly invested in your research project, deem you ready to graduate. The goal of the PhD is really to become the world's number-one expert in your own small piece of the huge biomedical landscape. The PhD student, his or her supervisor, usually the PI of the lab, and the thesis committee work together over several years to shape and direct the project, which is carried out by the student. Although many PhD programs involve some coursework, the focus from a very early stage is on the experiments.

While medical school, like graduate school in many other non-scientific disciplines, can be extremely expensive, people are generally actually *paid* to do PhDs in biomedical research labs. After a PhD is bestowed, the conventional trajectory is to do postdoctoral study, a "postdoc." This means that you work for a PI in his or her lab, but gradually take on more responsibility and ownership over your projects.

The ultimate goal of many researchers is a tenure-track faculty position. Tenure means job security; the university or college guarantees your employment as long as you don't do anything illegal or really stupid. At many schools, tenure is awarded when the promotion from assistant to associate professor occurs, often five years after your initial faculty appointment. Tenure is not a sure thing, and would-be associate professors are assessed based on their reputation among their peers, their service to the university, the papers they have published, and research grants they have been awarded.

The life of a PI is competitive and challenging. The budget of the National Institutes of Health (NIH), which funds most biomedical research, has not really grown much in the past twenty years or so. When I was doing my PhD, when Bill Clinton was president, about one in four NIH grant applications would get funded. Now it is closer to one in seven. This is not because the quality of the grant proposals is getting lower; in fact, the opposite is probably true. It is because there is not enough money available to support all the worthwhile research.

Publications are the currency of academic research science. As described in chapter 25, journals are ranked by impact factor, which is a measure of how many times papers published in that journal are cited. Publishing in high-impact journals increases a research scientist's clout and potential for funding and advancement. The problem is that the power that comes with even a single paper being published in an elite journal engenders a huge incentive to do whatever is necessary to get published. There is a direct correlation between journals' impact factors and the number of retractions, where a paper is removed from the publication record following identification of issues with methodology, ethics, or reproducibility. Highly publicized scandals, such as the retraction of Haruko Obokata's STAP cell papers (chapter 25), have brought these issues to the fore. A great deal of attention is now being paid to that matter of reproducibility. Online communities, universities, publishers, and funding bodies are changing the professional environment so that research standards are becoming significantly more stringent. More and more, we're seeing truly heartening examples of the inherent self-correcting nature of science.

On the other hand, there are certain factors currently limiting the advancement of science.

One major issue on every level is that people are just too busy. The PhD student or postdoc is juggling too many projects, and many feel pressure to cut corners, because the PI wants the results *now*, and often not just any results, but ones that confirm his or her hypothesis. Peer review, the critical assessment of other scientists' work, lies at the core of funding and publishing decisions. However, if the PI has too many grants and papers to review, he or she may delegate this work to less-experienced lab members without sufficient instruction or oversight. Further, in addition to teaching, sitting on committees, planning experiments, and going over data, the PI spends inordinate amounts of time writing and rewriting papers for publication and, especially, grant applications, the majority of which go unfunded.

So what is the solution?

Increase the NIH budget. Allow scientists to spend more time doing science and less time searching for funding. Increasing the availability of funding would also permit scientific departments to grow, providing more options for early-career scientists.

Who pays for the NIH budget? We do, with our taxes, as allocated by Congress. And what do you think happens when a member of Congress or the average nonscientist tries to read a basic science research paper in a biomedical journal? It might as well be written in a foreign language. Thus, outreach and communication to nonscientists is an essential responsibility of the researcher. The public at large must understand science and feel as though they are stakeholders in the research process. This is one of the key motivations that led me to write this book. It is my hope that increasing awareness and understanding of at least my own personal area of biomedical research will have a positive impact on its societal value and the drive to support science. Likewise, improved science education in our public schools is the foundation upon which our scientific success must be built. We can do better at every academic level, and we must.

How can we educate the public about science? Like many academic fields, science is inherently complex and contentious. If you ask two economists the same question, you often get diametrically opposed answers. The same can happen with judges looking at the specifics of a legal case. However, most other fields involve inductive reasoning and comparison to precedent, both of which are subject to interpretation.

Science is built on experimentation and quantification. Does this drug inhibit that protein or not? Will it reduce the spread of cancer cells better than the currently accepted treatment or not? Of course, the devil is in the details, and knowing precisely how an experiment was conducted is essential to appreciating whether the results can be trusted. Thus, even for experts, performing a full critical analysis of a piece of scientific work is a painstaking and time-consuming process.

The more that people know about the challenges and successes of basic research, the better it will be understood and valued. Similarly, people need to know how basic investigations are translated into clinical advances. Popular news media constantly bombards us with conflicting information on what to eat, what is safe, what drugs to take or avoid. Thus, while the onus is on the scientists to effectively communicate science to the public, people also need to be skeptical and not take everything "experts" say at face value. This problem, by the way, is probably unique to biomedical science. Explaining quantum physics to the public is even harder, but we don't have TV commercials telling us to ask our local physicist about what a particular subatomic particle can do for us.

That said, pseudoscience and claims without experimental basis are rampant. There are very few quick fixes, and if something seems too good to be true, it almost certainly is. Products such as dietary supplements and homeopathic "remedies" can be worse than just a waste of money if they prevent people from seeking proper medical care. Thus, regulatory bodies definitely need to become more stringent in creating and enforcing evidence-based criteria for would-be snake-oil salesmen.

It is my hope that this book has served to illuminate what I believe to be some of the more interesting and important advances both historical and at the cutting edge. I have focused primarily on areas in which I have some experience, knowledge, or expertise, so this has certainly not been a comprehensive exercise. However, a key facet to my perspective is that the use of microscopes to study cells underlies many of the most significant scientific breakthroughs regardless of the specific question or application. I have completely ignored critical topics extremely relevant to our future as a species, such as climate change and alternative energy. However, I hope you feel that this book has been interesting and useful, and I hope it can serve as a starting point, not an end, to your scientific inquiry, in particular, in the area of biomedical research. Funding for science must continue and grow, and the only way this will happen is if nonscientists understand the value of this work.

References and Sources
for Further Reading

Please look through the chapter-by-chapter references in the following section if you are motivated to find out more about the topics I have covered.

Introduction

Adams, D. *The Hitchhiker's Guide to the Galaxy*. London: Pan Books, 1979.

The Royal Swedish Academy of Sciences. Press Release. October 8, 2014.

Chapter 1—A Day the World Changed

Darwin, C. *On the Origin of Species*. John Murray, 1859.

Gest, H. "The discovery of microorganisms by Robert Hooke and Antoni van Leeuwenhoek, Fellows of The Royal Society." *Notes Rec. R. Soc. Lond.* 2004;58(2):187–201.

Hooke, R. *Micrographia*. BiblioBazaar, 1665.

Molyneux, T. Letter to the Secretary of the Royal Society from February 13, 1685, as quoted in Birch, T. *The History of the Royal Society of London*: Vol. 4, 1757.

Chapter 2—A Guided Tour of the Cell

Alberts, B, et al. *Molecular Biology of the Cell*, 4th edition. New York: Galand Science, 2002.

Archibald, JM. "Endosymbiosis and eukaryotic cell evolution." *Curr Biol.* 2015 Oct 5;25(19):R911–21.

Beck, J. Taking antibiotics can change the gut microbiome for up to a year. *The Atlantic*. November 16, 2015.

Blobel, G. "Protein targeting." Nobel Lecture. December 8, 1999.

Carmody, SR, and Wente, SR. "mRNA nuclear export at a glance." *J Cell Sci.* 2009 Jun 15;122(Pt 12):1933–7.

Coutinho, MF, et al. "Mannose-6-phosphate pathway: A review on its role in lysosomal function and dysfunction." *Mol Genet Metab.* 2012 Apr;105(4):542–50.

Dudek, J, et al. "Mitochondrial protein import: common principles and physiological networks." *Biochim Biophys Acta.* 2013 Feb;1833(2):274–85.

Grubb, JH, et al. "New strategies for enzyme replacement therapy for lysosomal storage diseases." *Rejuvenation Res.* 2010 Apr–Jun;13(2–3):229–36.

Lodish, H, et al. *Molecular Cell Biology*, 4th edition. New York: W.H. Freeman & Co., 2000.

Mitchell, P. "Coupling of phosphorylation to electron and hydrogen transfer by a chemi-osmotic type of mechanism." *Nature.* 1961 Jul 8;191:144–8.

Palade, GE, and Porter, KR. "Studies on the endoplasmic reticulum. I. Its identification in cells in situ." *J Exp Med.* 1954 Dec 1;100(6):641–56.

Palade, GE, and Siekevitz, P. "Liver microsomes; an integrated morphological and biochemical study." *J Biophys Biochem Cytol.* 1956 Mar 25;2(2):171–200.

Wente, SR, and Rout, MP. "The nuclear pore complex and nuclear transport." *Cold Spring Harb Perspect Biol.* 2010 Oct;2(10):a000562.

Chapter 3—The Central Dogma of Molecular Biology

Dabney, J, et al. "Complete mitochondrial genome sequence of a Middle Pleistocene cave bear reconstructed from ultrashort DNA fragments." *Proc Natl Acad Sci USA.* 2013 Sep 24;110(39):15758–63.

Llorente, MG, et al. "Ancient Ethiopian genome reveals extensive Eurasian admixture throughout the African continent." *Science.* 2015 Nov 13;350(6262):820–2.

Schmucker, D, et al. "Drosophila Dscam is an axon fuidance receptor exhibiting extraordinary molecular diversity." *Cell.* 2000 Jun 9;101(6):671–84.

Watson, JD. *The Double Helix: A Personal Account of the Discovery of the Structure of DNA.* New York: Atheneum, 1968.

Watson, JD, and Crick, FH. "Molecular structure of nucleic acids; a structure for deoxyribose nucleic acid." *Nature.* 1953 Apr 25;171(4356):737–8.

Chapter 4—Solving the Mystery of Life: The Road to the Double Helix

Avery, OT, et al. "Studies on the chemical nature of the substance inducing transformation of pneumococcal types. Inductions of transformation by a desoxyribonucleic acid fraction isolated from pneumococcus type III." *J Exp Med.* 1944 Feb 1;79(2):137–58.

Griffith, F. "The significance of pneumococcal types." *J Hyg* (Lond). 1928 Jan;27(2):113–59.

Hershey, AD, and Chase, M. "Independent functions of viral protein and nucleic acid in growth of bacteriophage." *J Gen Physiol.* 1952 May;36(1):39–56.

Judson, HF. *The Eighth Day of Creation: Makers of the Revolution in Biology.* New York: Touchstone Books, 1979.

Meselson, M, and Stahl, FW. "The replication of DNA in Escherichia coli." *Proc Natl Acad Sci USA.* 1958 Jul 15;44(7):671–82.

Watson, JD, and Crick, FH. "Molecular structure of nucleic acids; a structure for deoxyribose nucleic acid." *Nature.* 1953 Apr 25;171(4356):737–8.

Watson, JD, and Crick, FH. "Genetical implications of the structure of deoxyribonucleic acid." *Nature.* 1953 May 30;171(4361):964–7.

Chapter 5—Epigenetics: Beyond the Central Dogma

Dawkins, R. *The Selfish Gene.* Oxford: Oxford University Press, 1976.

Glausiusz, J. "Searching chromosomes for the legacy of trauma." *Nature News.* 11 June 2014.

Kellermann, NPF. "Epigenetic transmission of holocaust trauma: Can nightmares be inherited?" *Isr J Psychiatry Relat Sci.* 2013;50(1):33–9

Ravindran, S. "Barbara McClintock and the discovery of jumping genes." *Proc Natl Acad Sci USA.* 2012 Dec 11;109(50):20198–9.

Chapter 6—The Monk's Garden: Mendel's Law of Independent Assortment and How It Can Be Broken

Creighton, HB, and McClintock, BA. "Correlation of cytological and genetical crossing-over in Zea Mays." *Proc Natl Acad Sci USA.* 1931 Aug;17(8):492–7.

Hayden, EC. "Is the $1,000 genome for real?" *Nature News.* January 15, 2014.

Ott, J, et al. "Genetic linkage analysis in the age of whole-genome sequencing." *Nat Rev Genet.* 2015 May;16(5):275–84.

Chapter 7—The Revolutionary Reaction, or, How to Make DNA in Your Kitchen

Chien, A, et al. "Deoxyribonucleic acid polymerase from the extreme thermophile Thermus aquaticus." *J Bacteriol.* 1976 Sep;127(3):1550–7.

Mullis, K. "The polymerase chain reaction." Nobel Lecture. December 8, 1993.

Mullis, K, et al. "Specific enzymatic amplification of DNA in vitro: The polymerase chain reaction." *Cold Spring Harb Symp Quant Biol.* 1986;51(Pt 1):263–73.

Chapter 8—Piecing Together the Puzzle: How We Sequence DNA

Lander, ES, et al. "Initial sequencing and analysis of the human genome." *Nature*. 2001 Feb 15;409(6822):860–921.

Sanger, F. "The chemistry of insulin." Nobel Lecture. December 11, 1958.

Sanger, F. "Determination of nucleotide sequences in DNA." Nobel Lecture. December 8, 1980.

Sanger, F, et al. "Use of DNA polymerase I primed by a synthetic oligonucleotide to determine a nucleotide sequence in phage fl DNA." *Proc Natl Acad Sci USA*. 1973 Apr;70(4):1209–13.

Sanger, F, et al. "DNA sequencing with chain-terminating inhibitors." *Proc Natl Acad Sci USA*. 1977 Dec;74(12):5463–7.

Sanger, F, et al. "Nucleotide sequence of bacteriophage phi X174 DNA." *Nature*. 1977 Feb 24;265(5596):687–95.

Chapter 9—The Genome and Personalized Medicine: Progress, Promise, and Potential Problems

Flaherty, KT, et al. "Inhibition of mutated, activated BRAF in metastatic melanoma." *N Engl J Med*. 2010 Aug 26;363(9):809–19.

Nazarian R, et al. "Melanomas acquire resistance to B-RAF(V600E) inhibition by RTK or N-RAS upregulation." *Nature*. 2010 Dec 16;468(7326):973–7.

Zebisch, A, and Troppmair, J. "Back to the roots: The remarkable RAF oncogene story." *Cell Mol Life Sci*. 2006 Jun;63(11):1314–30.

Chapter 10—The Science, Technology, and Ethics of Manipulating the Genome

Hall, B, et al. "Overview: Generation of gene knockout mice." *Curr Protoc Cell Biol*. 2009 Sep;Chapter 19:Unit 19.12 19.12.1–17.

Liang, P, et al. "CRISPR/Cas9-mediated gene editing in human tripronuclear zygotes." *Protein Cell*. 2015 May;6(5):363–72.

Nathwani, AC. "Long-term safety and efficacy of factor IX gene therapy in hemophilia B." *N Engl J Med*. 2014 Nov 20;371(21):1994–2004.

Qasim, W, et al. "First clinical application of talen engineered universal CAR19 T cells in B-ALL." American Society of Hematology 57th Annual Meeting, 2015.

Stolberg, SG. "The biotech death of Jesse Gelsinger." *New York Times*. November 28, 1999.

Yang, L, et al. "Genome-wide inactivation of porcine endogenous retroviruses (PERVs)." *Science*. 2015 Nov 27;350(6264):1101–4.

Chapter 11—Science Fiction and Social Fiction: What Is and Is Not in Our Genes

Anderson, MD. "Even black preschool teachers are biased." *The Atlantic*. September 28, 2016.

Gilliam, WS, et al. "Do early educators' implicit biases regarding sex and race relate to behavior expectations and recommendations of preschool expulsions and suspensions?" *Yale University Child Study Center*. September 28, 2016.

Hunt-Grubbe, C. "The elementary DNA of Dr Watson." *Sunday Times*. October 14, 2007.

Herrnstein, R, and Murray, C. *The Bell Curve*. Free Press. 1994.

Lavy, V, and Sand, E. "On the origins of gender human capital gaps: short and long term consequences of teachers' stereotypical biases." National Bureau of Economic Research Working Paper 20909. 2015.

Chapter 12—The Jellyfish That Taught Us How to See

Avwioro, G. "Histochemical users of haematoxylin—A review." *Journal of Pharmacy and Clinical Sciences*. 2011 April–June; 1:24–34.

Baird, GS, et al. "Biochemistry, mutagenesis, and oligomerization of DsRed, a red fluorescent protein from coral." *Proc Natl Acad Sci USA*. 2000 Oct 24;97(22):11984–9.

Campbell, RE, et al. A monomeric red fluorescent protein. *Proc Natl Acad Sci USA*. 2002 Jun 11;99(12):7877–82.

Homer. *The Iliad*. New York: Penguin, 1998.

Huh, WK, et al. "Global analysis of protein localization in budding yeast." *Nature*. 2003 Oct 16;425(6959):686–91.

Matz, MV, et al. "Fluorescent proteins from nonbioluminescent Anthozoa species." *Nat Biotechnol*. 1999 Oct;17(10):969–73.

Patterson, GH, et al. "Use of the green fluorescent protein and its mutants in quantitative fluorescence microscopy." *Biophys J*. 1997 Nov;73(5):2782–90.

Smith, A. "The Nobel Prize in Chemistry 2008—Speed read." Nobelprize.org.

Stokes, GG. "On the change of refrangibility of light." *Phil. Trans. R. Soc. Lond.* 1852;142:463–562.

Chapter 13—How We See Clearly Inside Living Cells

Denk, W, and Svoboda, K. "Photon upmanship: Why multiphoton imaging is more than a gimmick." *Neuron*. 1997 Mar;18(3):351–7.

Frigault, MM, et al. "Live-cell microscopy—Tips and tools." *J Cell Sci*. 2009 Mar 15;122(Pt 6):753–67.

Galloway, JW. "The new microscopy." *New Scientist*. July 2, 1987.

Spiess, E, et al. "Two-photon excitation and emission spectra of the green fluorescent protein variants ECFP, EGFP and EYFP." *J Microsc*. 2005 Mar;217(Pt 3):200–4.

Chapter 14—Light-Sheet Microscopy, or, The Light in SPIM Stays Mainly in the Plane

Axelrod, D, et al. "Total internal inflection fluorescent microscopy." *J Microsc*. 1983 Jan;129(Pt 1):19–28.

Mattheyses, AL, et al. "Imaging with total internal reflection fluorescence microscopy for the cell biologist." *J Cell Sci*. 2010 Nov 1;123(Pt 21):3621–8.

Santi, PA. "Light sheet fluorescence microscopy: A review." *J Histochem Cytochem*. 2011 Feb;59(2):129–38.

Chapter 15—Super-Resolution Microscopy: Turning the Lights On One at a Time

Manley, S, et al. "High-density mapping of single-molecule trajectories with photoactivated localization microscopy." *Nat Methods*. 2008 Feb;5(2):155–7.

Rust, MJ, et al. "Sub-diffraction-limit imaging by stochastic optical reconstruction microscopy (STORM)." *Nat Methods*. 2006 Oct;3(10):793–5.

Thorley, JA, et al. "Super-resolution Microscopy: A Comparison of Commercially Available Options." Chapter 14 in *Fluorescence Microscopy: Super-Resolution and Other Novel Techniques*. Edited by Cornea, A, and Conn, PM. Amsterdam: Elsevier Science, 2014.

Chapter 16—What Makes the Glowworm Glow? The Advantages of Luminescent Imaging

Cook, SH, and Griffin, DE. "Luciferase imaging of a neurotropic viral infection in intact animals." *J Virol*. 2003 May;77(9):5333–8.

Gupta, P, et al. "Metastasis of breast tumor cells to brain is suppressed by phenethyl isothiocyanate in a novel in vivo metastasis model." *PLoS One*. 2013 Jun 27;8(6):e67278. Print 2013.

Chapter 17—More Ways to Take Pretty and Enlightening Pictures

Axelrod, D. "Lateral motion of fluorescently labeled acetylcholine receptors in membranes of developing muscle fibers." *Proc Natl Acad Sci USA*. 1976 Dec;73(12):4594–8.

Axelrod, D, et al. "Mobility measurement by analysis of fluorescence photobleaching recovery kinetics." *Biophys J*. 1976 Sep;16(9):1055–69.

Feinberg, EH, et al. "GFP Reconstitution Across Synaptic Partners (GRASP) defines cell contacts and synapses in living nervous systems." *Neuron.* 2008 Feb 7;57(3):353–63.

Forster, T. "Transfer mechanisms of electronic excitation energy." *Radiation Research Supplement.* 1960;2:326–39.

Fussner, E, et al. "Open and closed domains in the mouse genome are configured as 10-nm chromatin fibres." *EMBO Rep.* 2012 Nov 6;13(11):992–6.

Grynkiewicz, G, et al. "A new generation of Ca2+ indicators with greatly improved fluorescence properties." *J Biol Chem.* 1985 Mar 25;260(6):3440–50.

Hu, CD, et al. "Visualization of interactions among bZIP and Rel family proteins in living cells using bimolecular fluorescence complementation." *Mol Cell.* 2002 Apr;9(4):789–98.

Jankowski, A, et al. "In situ measurements of the pH of mammalian peroxisomes using the fluorescent protein pHluorin." *J Biol Chem.* 2001 Dec 28;276(52):48748–53.

Koppel, DE, et al. "Dynamics of fluorescence marker concentration as a probe of mobility." *Biophys J.* 1976 Nov;16(11):1315–29.

Macpherson, LJ, et al. "Dynamic labelling of neural connections in multiple colours by trans-synaptic fluorescence complementation." *Nat Commun.* 2015 Dec 4;6:10024.

Nagai, T, et al. "Circularly permuted green fluorescent proteins engineered to sense Ca2+." *Proc Natl Acad Sci USA.* 2001 Mar 13;98(6):3197–202.

Pologruto, TA, et al. "Monitoring neural activity and [Ca2+] with genetically encoded Ca2+ indicators." *J Neurosci.* 2004 Oct 27;24(43):9572–9.

Schlessinger, J, et al. "Lateral transport on cell membranes: Mobility of concanavalin A receptors on myoblasts." *Proc Natl Acad Sci USA.* 1976 Jul;73(7):2409–13.

Urban, E, et al. "Electron tomography reveals unbranched networks of actin filaments in lamellipodia." *Nat Cell Biol.* 2010 May;12(5):429–35.

Chapter 18—How Cells Die

Barber, DL, et al. "Restoring function in exhausted CD8 T cells during chronic viral infection." *Nature.* 2006 Feb 9;439(7077):682–7.

Brouckaert, G, et al. "Phagocytosis of necrotic cells by macrophages is phosphatidylserine dependent and does not induce inflammatory cytokine production." *Mol Biol Cell.* 2004 Mar;15(3):1089–100.

Dolan, DE, and Gupta, S. "PD-1 pathway inhibitors: Changing the landscape of cancer immunotherapy." *Cancer Control.* 2014 Jul;21(3):231–7.

Gorman, C. "Medicine Nobel recognizes 'self-eating' cells." *Scientific American*. October 3, 2016.

Knudson, AG. "Mutation and cancer: Statistical study of retinoblastoma." *Proc Natl Acad Sci USA*. 1971 Apr;68(4):820–3.

Takeshige, K, et al. "Autophagy in yeast demonstrated with proteinase deficient mutants and conditions for its induction." *Journal of Cell Biology*. 1992;119:301–311.

Snell, GD. "Studies in histocompatibility." Nobel Lecture. December 8, 1980.

Chapter 19—The Mystery of HIV

Burtey, A, et al. "Dynamic interaction of HIV1 Nef with the clathrin mediated endocytic pathway at the plasma membrane." *Traffic*. 2007 Jan;8(1):6176.

Daecke, J, et al. "Involvement of clathrin-mediated endocytosis in human immunodeficiency virus type 1 entry." *J Virol*. 2005 Feb;79(3):1581–94.

Miyauchi, K, et al. "HIV enters cells via endocytosis and dynamin-dependent fusion with endosomes." *Cell*. 2009 May 1;137(3):433–44.

Uchil, PD, and Mothes, W. "HIV entry revisited." *Cell*. 2009 May 1;137(3):402–4.

Yeganeh, B, et al. "Suppression of influenza A virus replication in human lung epithelial cells by noncytotoxic concentrations bafilomycin A1." *Am J Physiol Lung Cell Mol Physiol*. 2015 Feb 1;308(3):L270–86.

Chapter 20—What Are Organs and Why Do We Have Them?

Abercrombie, M. "Contact inhibition and malignancy." *Nature*. 1979 Sep 27;281(5729):259–62.

Brownlee, C. "Biography of Barry S. Coller." *Proc Natl Acad Sci USA*. 2004 Sept;101(36):13111–3.

Califf, R, et al. "Use of a monoclonal antibody directed against the platelet glycoprotein IIb/IIIa receptor in high-risk coronary angioplasty. The EPIC investigation." *N Engl J Med*. 1994 Apr 7;330(14):956–61.

Dhiren, S. "Coronary artery bypass grafting (CABG): Past present and future." *Gujrat Medical Journal*. 2010 Jul;65(2):82–7.

Erbel, R, and Eggebrecht, H. "Aortic dimensions and the risk of dissection." *Heart*. 2006 Jan;92(1):137–42.

Gupta, A, et al. "Coronary arterial revascularization: Past, present, future: Part I—historical trials." *Clin Cardiol*. 2006 Jul;29(7):290–4.

Chapter 21—The Kidney: Cells in Concert

Nielsen, S, et al. "Vasopressin increases water permeability of kidney collecting duct by inducing translocation of aquaporin-CD water channels to plasma membrane." *Proc Natl Acad Sci USA*. 1995 Feb 14;92(4):1013–7.

Preston, GM, and Agre, P. "Isolation of the cDNA for erythrocyte integral membrane protein of 28 kilodaltons: Member of an ancient channel family." *Proc Natl Acad Sci USA*. 1991 Dec 15;88(24):11110–4.

Roy, A, et al. "Collecting duct intercalated cell function and regulation." *Clin J Am Soc Nephrol*. 2015 Feb 6;10(2):305–24.

Chapter 22—The Brain: How It Works, How We Learn, and Why It's Hard to Study

Castellucci, V, et al. "Neuronal mechanisms of habituation and dishabituation of the gill-withdrawal reflex in Aplysia." *Science*. 1970 Mar 27;167(3926):1745–8.

Coyle, D. "How to grow a SuperAthlete." *New York Times*. March 4, 2007.

Doidge, N. *The Brain That Changes Itself*. New York: Penguin Books, 2007.

Hodgkin, AL, and Huxley, AF. "Action potentials recorded from inside a nerve fibre." *Nature*. 1939 Oct 21;144:710–1.

Love, S. "Demyelinating diseases." *J Clin Pathol*. 2006 Nov;59(11):1151–9.

McKenzie, IA, et al. "Motor skill learning requires active central myelination." *Science*. 2014 Oct 17;346(6207):318–22.

Richardson, DS, and Lichtman, JW. "Clarifying tissue clearing." *Cell*. 2015 Jul 16;162(2):246–57.

Roberts, AC, and Glanzman, DL. "Learning in Aplysia: Looking at synaptic plasticity from both sides." *Trends Neurosci*. 2003 Dec;26(12):662–70.

Schwiening, CJ. "A brief historical perspective: Hodgkin and Huxley." *J Physiol*. 2012 Jun 1;590(Pt 11):2571–5.

Tamariz, E, and Varela-Echavarría, A. "The discovery of the growth cone and its influence on the study of axon guidance." *Front Neuroanat*. 2015 May 15;9:51.

Taylor, AP. "Myelin's role in motor learning." *The Scientist*. October 16, 2014.

Chapter 23—The Immune System: How It Defends Us and Sometimes Attacks Us

Barrios, C, et al. "Gut-microbiota-metabolite axis in early renal function decline." *PLoS One*. 2015 Aug 4;10(8):e0134311.

Beck, J. "Taking antibiotics can change the gut microbiome for up to a year." *The Atlantic*. November 16, 2015.

Biswas, M, et al. "Treg: Tolerance vs immunity." *Oncotarget*. 2015;6(24):19956–7.

Dunkelberger, JR, and Wen-Chao Song, WC. "Complement and its role in innate and adaptive immune responses." *Cell Res*. 2010;20:34–50.

Elert, E. "Calling cells to arms." *Nature*. 2013 Dec 19;504(7480):S2–3.

Goodrich, JK, et al. "Human genetics shape the gut microbiome." *Cell*. 2014 Nov 6;159(4):789–99.

Goodrich, JK, et al. "Genetic determinants of the gut microbiome in UK twins." *Cell Host Microbe*. 2016 May 11;19(5):731–43.

Gorman, C. "Cancer immunotherapy: The cutting edge gets sharper." *Scientific American*. October 1, 2015.

Gratz, IK, et al. "The life of regulatory T cells." *Annals of the New York Academy of Sciences*. 2013;1283:8–12.

Jackson, MA, et al. "Proton pump inhibitors alter the composition of the gut microbiota." *Gut*. 2016 May;65(5):749–56.

König, M, et al. "Tregalizumab—A monoclonal antibody to target regulatory T cells." *Front Immunol*. 2016 Jan 25;7:11.

McGinley, L. "The list of cancers that can be treated by immunotherapy keeps growing." *Washington Post*. April 19, 2016.

Newkirk, VR. "Cancer: The final frontier." *The Atlantic*. April 21, 2016.

Vanamala, JK, et al. "Can your microbiome tell you what to eat?" *Cell Metab*. 2015 Dec 1;22(6):960–1.

Weintraub, K. "Releasing the breaks." *Nature*. 2013 Dec 19:504(7480):S6–8.

Xudong, L, and Zheng, Y. "Regulatory T cell identity: Formation and maintenance." *Trends in Immun*. 2015 June;36(6):344–353.

Zaura, E, et al. "Same exposure but two radically different responses to antibiotics: Resilience of the salivary microbiome versus long-term microbial shifts in feces." *MBio*. 2015 Nov 10;6(6):e01693-15.

Chapter 24—How We Amost Cured Cancer

Flamme, I, et al. "Molecular mechanisms of vasculogenesis and embryonic angiogenesis." *J Cell Physiol*. 1997 Nov;173(2):206–10.

Folkman, J. "Tumor angiogenesis: Therapeutic implications." *N Engl J Med*. 1971 Nov 18;285(21):1182–6.

Folkman, J. "Is angiogenesis an organizing principle in biology and medicine?" *J Pediatr Surg*. 2007 Jan;42(1):1–11.

Grady, D. "A failure to verify a cancer advance is raising concern." *New York Times*. November 13, 1998.

Kolata, G. "Hope in the lab: A special report; A cautious awe greets drugs that eradicate tumors in mice." *New York Times*. May 3, 1998.

Ma, J, and Waxman, DJ. "Combination of anti-angiogenesis with chemotherapy for more effective cancer treatment." *Mol Cancer Ther*. 2008 December;7(12):3670–84.

"Nobel prize winner denies cancer quote." *LA Times*. May 07, 1998. From Reuters.

Ryan, DP, et al. "Reality testing in cancer treatment: The phase I trial of endostatin." *Oncologist*. 1999;4(6):501–8.

Chapter 25—Ethics, Ambition, and the Greatest Discovery That Wasn't

De Los Angeles, A, et al. "Failure to replicate the STAP cell phenomenon." *Nature*. 2015 Sep 24;525(7570):E6–9.

Goodyear, D. "The stress test." *The New Yorker*. 2016 Feb 29.

Green, H. "The birth of therapy with cultured cells." *Bioessays*. 2008 Sep;30(9):897–903.

Grynkiewicz, G, et al. "A new generation of Ca2+ indicators with greatly improved fluorescence properties." *J Biol Chem*. 1985 Mar 25;260(6):3440–50.

Konno, D, et al. "STAP cells are derived from ES cells." *Nature*. 2015 Sep 24;525(7570):E4–5.

Obokata, H, et al. "Stimulus-triggered fate conversion of somatic cells into pluripotency." *Nature*. 2014 Jan 30;505(7485):641–7. (retracted)

Obokata, H, et al. "Bidirectional developmental potential in reprogrammed cells with acquired pluripotency." *Nature*. 2014 Jan 30;505(7485):676–80. (retracted)

Takahashi, K, and Yamanaka, S. "Induction of pluripotent stem cells from mouse embryonic and adult fibroblast cultures by defined factors." *Cell*. 2006 Aug 25;126(4):663–76.

Tang, MK, et al. "Transient acid treatment cannot induce neonatal somatic cells to become pluripotent stem cells." *F1000Res*. 2014 May 8;3:102.

"STAP paper co-author Sasai commits suicide." *Japan Times*. August 5, 2014.

Chapter 26—The Nano Revolution

Berger, A. "How does it work? Magnetic resonance imaging." *BMJ*. 2002 Jan 5;324:35.

Biju, V, et al. "Nanoparticles speckled by ready-to-conjugate lanthanide complexes for multimodal imaging." *Nanoscale*. 2015 Sep 28;7(36):14829–37.

Crichton, M. *Prey*. New York: Harper, 2002.

Drexler, E. *Engines of Creation*. New York: Anchor, 1986.

Nel, AE, et al. "Understanding biophysicochemical interactions at the nano-bio interface." *Nat Mater*. 2009 Jul;8(7):543–57.

Osborne, L. "The gray goo problem." *New York Times*. December 14, 2003.

Salvati, A, et al. "Transferrin-functionalized nanoparticles lose their targeting capabilities when a biomolecule corona adsorbs on the surface." *Nat Nanotechnol*. 2013 Feb;8(2):137–43.

Chapter 27—Money, Power, Ambition, and the Pursuit of Knowledge: Thoughts on the Current State of Science

Crichton, M. *Five Patients*. London: Arrow/Children's (a Division of Random House), 1995.

Donati, G. "Careers in academia: Different options." Nature Jobs Blog. October 9, 2015.

Fang, FC, and Casadevall, A. "Retracted science and the retraction index." *Infect Immun*. 2011 Oct;79(10):3855–9.

Goodman, SN, et al. "What does research reproducibility mean?" *Science Translational Medicine*. 2016 June; Vol 8(341).

Lehrer, J. "The truth wears off." *The New Yorker*. 2010 Dec 13.

Nurse, P. "US biomedical research under siege." *Cell*. 2006 Jan 13;124(1):9–12.

Schillebeeck, M, et al. "The missing piece to changing the university culture." *Nat Biotechnol*. 2013 Oct;31(10):938–41.

"New and early stage investigator policies." NIH Office of Extramural Research. February 10, 2014.

Notice Number: NOTOD15048. Ruth L. Kirschstein National Research Service Award (NRSA) Stipends, Tuition/Fees and Other Budgetary Levels Effective for Fiscal Year. Release Date: December 30, 2014. Issued by National Institutes of Health (NIH) Agency for Healthcare Research and Quality (AHRQ) Health Resources Services Administration (HRSA).

Research Project Success Rates by Type and Activity. NIH Research Portfolio Online Reporting Tools (RePORT).

Tenured/Tenure-Track Faculty Salaries. Results of the 2012–13 Faculty in Higher Education Salary Survey by Discipline, Rank and Tenure Status in Four-Year Colleges and Universities conducted by The College and University Professional Association for Human Resources (CUPA-HR). HigherEdJobs.com.

Index

Acknowledgments

First, I would like to thank those researchers and educators who went out of their way to help make me the scientist, and person, I am today. In addition to the people described in the Dedication (Don Jackson, Ruth Abramson, and Sandy Simon), these include (but are not limited to): Sankar Sengupta, Mark Unno, Diomedes Logothetis, Terry Krulwich, Scott Henderson, Pat Wilson, Mike Lipkowitz, Sandy Masur, Michael Tibbetts, Sandy Schmid, Frances Brodsky, John Heath, Volodya Gelfand, Bob Goldman, and Phil Hockberger.

Whenever I have read an author's statement at the end of a book that it "takes a team" to successfully bring a project to completion, I have always felt that he or she was being humble and/or was lucky to have a group of eager assistants at the ready. Now that I have been through this process, I see that the number of favors the author must request, as well as opinions and potential alternative perspectives, really does make one realize the importance of friends, family members, and colleagues. The patience and helpfulness of quite a large number of people have made writing this considerably easier.

Without the support and guidance of my agent, John Willig, from Literary Services, Inc., this book would never have gotten off the ground. Similarly, from the initial conception of this project through completion, the team at BenBella Books, especially Glenn Yeffeth, Leah Wilson, Vy Tran, and Sarah Dombrowsky, has been absolutely instrumental in every aspect of this book. In particular, my editor

David Bessmer deserves a tremendous amount of appreciation—his assistance and helpful suggestions improved this book immeasurably.

I also want to specifically thank the following people for their help reading drafts of the work in progress and giving me very useful feedback: Maddy Parsons, Abdullah and Sara Khan, Jeff Moskin, Joan Rosenfeld, and Jim Rappoport. Finally, I want to thank my parents, my children, and especially my wife, Ema, for their patience, love, and support throughout this process.

About the Author

DR. JOSHUA Z. RAPPOPORT received a bachelor's degree in biology from Brown University and a PhD from the Program in Mechanisms of Disease and Therapeutics at the Mount Sinai School of Medicine Graduate School of Biological Sciences of New York University. Following defense of his thesis, Dr. Rappoport performed postdoctoral work at the Rockefeller University in New York City in the Laboratory of Cellular Biophysics. Subsequently, he was recruited as a tenured faculty member in the School of Biosciences at the University of Birmingham in England.

In 2014, Dr. Rappoport returned to the United States, where he is currently the director of the Center for Advanced Microscopy and Nikon Imaging Center at the Northwestern University Feinberg School of Medicine and a faculty member in the Department of Molecular and Cell Biology. Dr. Rappoport's scientific perspective aims to develop and apply cutting-edge microscopy to address fundamental questions in cell biology.

Dr. Rappoport lives in Chicago with his wife, Ema, a neuroscientist, and their dog, Kris.